NATURAL GAS: THE NEW ENERGY LEADER

By Ernest J. Oppenheimer, Ph.D

NATURAL GAS: THE NEW ENERGY LEADER

By Ernest J. Oppenheimer, Ph.D.

Second Printing, July 1981
© Copyright 1980, 1981 by Ernest J. Oppenheimer, Ph.D.

Publisher: Pen and Podium Productions
40 Central Park South
New York, New York 10019

Printed in the United States of America

ISBN-0-9603982-2-8

Library of Congress Catalog Card No. 80-83793

Introduction to the Second Printing

When writing this book last year, I expressed optimism about the future of deep gas (see Chapter 4). I would like to share with the readers my assessment of the latest happenings in this important area.

The U.S. is experiencing a deep gas boom. In 1980 a record 859 wells 15,000 or more feet deep were drilled at a cost of $4 billion (*Petroleum Engineer International,* March 1981). The pace of exploration for deep gas has accelerated in 1981, with activity in key regions running as much as 50 percent ahead of last year.

The deep gas boom rests on a solid foundation of plentiful resources, sophisticated technological achievements, and favorable economic realities.

Deep gas accumulations, equaling or surpassing in hydrocarbon energy content the biggest U.S. oil fields, have already been discovered. The Tuscaloosa Trend in Louisiana is estimated to contain 60 trillion cubic feet of gas, the equivalent of 10 billion barrels of oil. The Anadarko Basin in Oklahoma and Texas may hold as much as 180 trillion cubic feet of deep gas, equaling 30 billion barrels of oil. Estimates for deep gas deposits in the Rocky Mountain Overthrust Belt range as high as 200 to 400 trillion cubic feet. The exploration for deep gas in other areas is imminent, notably in the Eastern Overthrust Belt, situated within the energy-starved region stretching from New England to Alabama.

The industries serving the exploration and drilling companies have done a magnificent job in overcoming the obstacles involved in deep earth penetrations. New steel alloys were developed to produce pipe that could withstand the high temperatures and pressures, as well as the corrosive substances encountered at great depths. Seismic technology has advanced

3

to the point where the success rate of finding deep gas compares favorably with that of locating hydrocarbons in shallow areas. High-strength proppants have been developed to keep the gas flowing from deep wells that might otherwise be closed down by the tremendous pressures of the earth.

Drilling for deep gas is expensive. In 1980 the average cost for drilling and completing a deep well was $4.56 million; this figure is almost twenty times the average cost of shallow wells. Because only about half of the deep wells drilled produce commercial quantities of gas, the cost per productive well was around $9 million. The federal government took cognizance of these realities; the Natural Gas Policy Act of 1978 deregulated the price of gas produced from below 15,000 feet.

While the costs are high, deep gas wells have a number of redeeming features: (1) speedy realization of commercial results; (2) high rates of production; (3) long life resulting from large reserves; (4) primary recovery of 70–80 percent of the gas in place; (5) minimal processing prior to end use; and (6) location near existing pipe lines or markets.

Under favorable conditions, a deep gas well can be drilled, completed, and put into production in about a year. As a result, it can contribute to income relatively soon after the investment has been made. In contrast, many alternative energy projects require years of preparation and construction, tying up capital for long periods before yielding any income (or energy).

The average deep well produces substantially more gas than the average shallow well. Production from deep wells is often prolific. Wells producing several million cubic feet of gas a day are not uncommon. At unregulated prices, the invested capital can be quickly recovered, if everything goes well.

The high rate of daily production is combined with long life made possible by the huge reserves being tapped. An executive of the GHK Company, which is active in the Anadarko Basin,

4

noted that according to their calculations some of the deep wells can sustain high rates of production for 20 to 30 years. Gas pipe line operators and utilities understandably give preference to such assured long-term sources of supply.

Deep gas is under great pressure, stimulating speedy and high recovery rates. 70–80 percent of the gas in place is likely to be recovered without external stimulation. In contrast, primary recovery from oil wells averages only 25–30%.

Deep gas comes in two varieties: sweet and sour. Sweet gas can be used essentially as it comes out of the earth, with minimum processing. Sour gas contains sulfur, which must be removed before the gas is allowed to enter the pipe lines. While the removal of sulfur adds to the cost of production, this procedure is not nearly as expensive as refining crude oil into such end products as gasoline or fuel oil.

Much of the deep gas that has been found thus far is situated near existing pipe lines. The Anadarko Basin in Oklahoma and Texas and the Tuscaloosa Trend in Louisiana are well supplied with pipe lines that have hitherto served to transport gas from shallower locations. If large quantities of gas are discovered in the Eastern Overthrust Belt, the closeness to markets will provide significant cost benefits.

These economic advantages, along with the decontrolled price, are providing the incentives for the record development of U.S. deep gas resources. The large-scale drilling activity is adding substantial amounts of natural gas energy to current supplies, as well as greatly increasing reserves. Deep gas is emerging as the most important new source of domestic energy for decades to come. For the first time in years, the U.S. can look forward to a prolonged period of energy abundance from its own resources. If present trends continue, natural gas will become the new energy leader much sooner than I had anticipated.

July 1981   Ernest J. Oppenheimer, Ph.D.

Books by Ernest J. Oppenheimer, Ph.D.

Natural Gas: The New Energy Leader.
A Realistic Approach to U.S. Energy Independence.
What You Should Know About Inflation.
The Inflation Swindle.

Contents

Foreword

By Dr. Hollis D. Hedberg, Professor of Geology (emeritus),
Princeton University:
I have been asked by the author to write a Foreword to this
book. I am glad to oblige because I believe Dr. Oppenheimer's
book performs a valuable service in presenting a clear and
comprehensive account of methane gas and its importance to
this country. He writes in a readable form which can be readily
understood by the general public. The book should have a
salutary influence on national thinking about the role of natural
gas.

Dr. Oppenheimer is rightly impressed with the merits of
methane gas as a source of energy and looks forward enthusias-
tically to its increasing utilization and rapidly growing role in
supplementing and replacing oil as a fuel. He is a social scientist
and not a professional geologist or engineer, yet he has success-
fully woven geology and technology into an informative and
interesting account. His geological comments appear sound and

9

are commonly backed adequately by literature references and quotations. I agree with his conclusion that methane gas is more abundant on this earth than is commonly recognized.

The author discusses at length the economic aspects of gas production and marketing. He correctly notes the harmful consequences of artificial pricing by government fiat and the burdening effects of overregulation on the natural gas industry.

Dr. Oppenheimer is impressed with the ability of the gas industry's leadership and its dedication to the best interests of the country. He emphasizes the need for government policies that will promote rather than hinder the development of the plentiful methane gas resources available in the U.S.

Anyone concerned about the national energy situation will profit from reading this stimulating book.

Preface

About two years ago I started intensive research on the energy problem. My initial attention was focused primarily on oil. Gas was frequently mentioned as a byproduct of oil production. In the fall of 1979 I attended the Conference on Long-Term Energy Resources in Montreal, which was sponsored by the United Nations Institute for Training and Research (UNITAR). Many scientific papers presented at that conference dealt with gas. The authors of these studies emphasized that gas was an important fuel in its own right, and not merely an adjunct to oil.

The more I looked into gas, the more impressed I was with its unique qualities and its great potential. I liked its cleanliness, its versatility, its abundance, and its relative cheapness. I was intrigued with some of the ways gas was formed, particularly the generation of methane by anaerobic microorganisms. The fact that oil is converted into methane (gas) at high temperatures and that methane is the most stable of all hydrocarbons added to the allure. It occurred to me that viewed objectively, gas may be more important than oil in the overall energy scheme.

My analysis has led me to the conclusion that natural gas may be poised for the biggest growth in its history. During the next two decades gas has the potential of becoming the leading fuel in the U.S.A. Gas will improve the environment, strengthen the domestic economy, and lessen dependence on imported oil.

These laudable goals can be achieved only with the cooperation of federal, state, and local governments. Hitherto, the gas industry has been overregulated and underutilized. That approach is inappropriate while the country is trying to achieve a greater measure of energy independence.

The U.S.A. has plentiful gas resources. The shortage of gas that occurred in the 1970's was caused by unsound government pricing policies, not by lack of gas in the ground. By utilizing conventional and unconventional sources, including gasification of coal, peat, and oil shale, U.S. gas supplies should last for centuries. Moreover, the conversion of plant materials into gas holds the promise of permanent energy abundance.

Throughout this book the terms "natural gas" and "methane" are used interchangeably. Methane is the principal ingredient of natural gas. The methane molecule consists of one carbon atom and four hydrogen atoms (CH_4). My book is dedicated to the understanding of this molecule and the important role it can play in our destiny.

I have tried to keep to a minimum the use of uncommon terms or technical language. However, in a book on natural gas some words or phrases not used in everyday language will occur. Wherever possible, I have tried to explain the meaning of these terms when they were first used. In addition, a Glossary defining such terms is presented in the back of the book.

October 1980

Ernest J. Oppenheimer, Ph.D.

Sources

Many sources have contributed to the realization of this book. My files of research materials on natural gas are filled with reports and studies by the American Gas Association, the Gas Research Institute, and the Institute for Gas Technology. I also have a complete set of papers presented at the Montreal Conference on Long-Term Energy Resources, sponsored by the United Nations Institute for Training and Research (UNITAR). I attended that conference in the fall of 1979. Much of the material used in the preparation of my previous energy book *A Realistic Approach to U.S. Energy Independence* was also relevant to the present work.

I like to take this opportunity to thank a number of individuals for their valuable counsel and helpfulness. Mr. George H. Lawrence, president of the American Gas Association, has encouraged me in writing this book and has facilitated my meeting with gas industry leaders. His staff includes many individuals with exceptional know-how. I shall single out two staff members for special mention because I called on them for assistance most frequently: Mr. Robert B. Kalisch, Director, Gas Supply and Statistics; and Mr. Michael I. German of the Policy Evaluation and Analysis Group. In addition to making his own perceptive comments, Mr. German coordinated the ideas presented by his colleagues.

Mr. Harold S. Walker, Jr., Executive Director of the New York Gas Group, was most helpful in contributing a broad perspective on the role of gas. He also gave me the benefit of his background as an English professor, carefully checking the manuscript for language usage and grammar.

A number of gas company executives provided specific information as well as invaluable judgment based on many years of experience. I had the privilege of interviewing the following industry leaders: Mr. Robert H. Willis, Chairman and President, Connecticut Natural Gas Company; Mr. Joseph R. Rensch, President, Pacific Lighting Corporation; Mr. O.C. Davis, Chairman, Peoples Energy Corporation; Mr. B.Z. Kastler, Chairman and President, Mountain Fuel Supply Company; and Mr. Eugene H. Luntey, President, Brooklyn Union Gas Company.

The primary focus of this book is on current and future developments. Because my informants are industry leaders and decision makers, their orientation is likely to have an impact on events. I believe the overall effect of these contacts has been to guide me in the direction of realism.

While I am indebted to others for their contributions, the ultimate responsibility for putting this book together rests on me. I found the task challenging. I hope the readers will gain as much satisfaction and intellectual stimulation reading the book as I had writing it.

14

1. Entering the Gas Age

For most of the twentieth century, oil has been the dominant fuel in the U.S.A. The events of the past several years have seriously eroded confidence in oil and have laid the foundation for the emergence of new energy leadership. On the basis of merit, natural gas is the most likely candidate to replace oil as the leading source of energy.

Gas is an ideal fuel. It is clean; it is plentiful; it is versatile; it is easy to transport and to store; it is relatively inexpensive.

Gas is the cleanest of all fossil fuels. It is reassuring to know that the increased use of gas will have beneficial effects on the environment. The more gas is used to replace oil, the cleaner the air will be.

The known gas resources in the U.S.A. are greater than the known oil resources. Moreover, it is likely that large additional quantities of gas will be found in deep locations (below 15,000 feet) and in unexplored territories largely owned by the federal government, including offshore areas. Unconventional sources of gas, such as western tight sands, Devonian shale, methane

15

in coal seams, and geopressured reservoirs, could add significantly to future supplies. Similarly, gasification of coal, peat, and oil shale could provide methane for many decades. Finally, gas can be obtained from renewable sources, including landfills, organic waste products, and plants grown on land and in oceans. As long as the sun continues to shine, plants will absorb solar energy by means of photosynthesis, and anaerobic microorganisms (able to live in the absence of oxygen) will eventually transform the plants into methane. These natural processes can be utilized to supply human gas energy requirements until the end of time.

Gas is a versatile fuel. It can be used to heat homes and offices. It can provide power to industrial plants and utilities. It can fuel automobiles and trucks. It can provide feedstock to chemical plants for the manufacture of plastics. It can do virtually all of the things oil can do. Moreover, in many applications it can perform the job better and cheaper than oil.

The million mile pipeline network for transporting gas, and the many regional and local storage facilities, give gas significant advantages over other fuels. Gas transports easily and cheaply. It keeps well in storage. It is always ready to do its job on instant demand. This reality is perhaps best illustrated by the common experience of turning on the gas range and having an immediate flame to cook a meal.

The economic advantages of gas over oil are impressive. The rate of primary recovery from gas wells averages 70%, compared with less than 30% for oil wells. Gas is ready to be used by consumers essentially right out of the pipeline; crude oil has to be refined before it can be utilized. Gas is so clean that the need for pollution control equipment is greatly reduced. From wellhead to consumer, pipeline transportation of gas is unexcelled in efficiency. The availability of gas supplies predominantly from domestic sources avoids the high risks and external costs of imported oil.

Gas has been used as a fuel for a long time. However, throughout its modern history gas has been overshadowed by oil. Its linkage with oil has prevented gas from the realization of its full potential. For example, until recently drilling for gas was an adjunct to drilling for oil. The latter cannot exist beyond certain temperatures. Therefore, virtually no drilling took place in locations having such unfavorable characteristics for finding oil. However, gas can exist at much higher temperatures than oil. Actually, some of the giant gas fields at great depths may have resulted partly from the heat-induced conversion of oil to gas. Drilling for deep gas had to await developments that made it economically attractive to drill for gas in terms of its own merits, rather than as an adjunct to oil. The emergence of commercial drilling for deep gas on a significant scale in the late 1970's may be considered the beginning of a new era, one in which gas is emerging as the dominant fuel.

The current trend of pipeline and distribution companies increasing their gas supply operations will facilitate the new role of gas as energy leader. By assuring long-term supplies of gas from a variety of captive sources, the gas industry will be in an increasingly sound position to take care of its customers.

The new era is already beginning to change the nature and philosophy of managements. Gas companies are attracting some of the most dynamic business executives, as well as top graduates of business and engineering schools. The investment community may be starting to recognize that the gas business is in the early stages of again becoming a growth industry. This recognition should facilitate the realization of the ambitious capital expenditure programs by gas companies over the next two decades.

To demonstrate its energy leadership, gas must help solve the current energy problem by providing increased supplies to replace imported oil. Gas is superior to oil in most stationary applications, such as providing heat and power to homes, offices, factories, and utilities. As a matter of national policy, gas should be given every encouragement to replace oil in those applications. If this were done, the oil import bill would be considerably reduced.

Unsound government policies have been the main obstacles to the emergence of gas as the new dominant energy source. Controls over prices and marketing, and many other regulatory constraints, have been serious handicaps. However, an increasing number of government officials are beginning to recognize the vital contributions that gas can make to the economy and to the national interest. As this orientation gains increased acceptance, it will facilitate the onset of the gas age.

Future historians will probably look back upon the energy crisis of the 1970's as the spawning ground for the emergence of gas as the new energy leader. The American people should welcome the gas age with enthusiasm, for it will help usher in a period of greater prosperity, a cleaner environment, and increased energy independence.

2. The Origins of Methane (Natural Gas)

In getting acquainted with methane, it is important to have some understanding of its origins. The science of geology has devoted much attention to this subject. There is broad agreement among scientists that methane originates from the following sources: (1) Organic matter in sediments whose decomposition is being promoted by heat; (2) The action of microorganisms that convert organic debris into methane; (3) The conversion of oil and other heavy hydrocarbons into methane at high temperatures; and (4) Coal which releases methane as it matures. In addition, some scientists have presented the hypothesis that methane may be part of our planet's original makeup and may be present in the interior layers of the earth, from where it migrates to the surface. This hypothesis remains to be proven and is not generally accepted.

Methane has been described as "the gaseous phase of petroleum."[1] It is likely to be found wherever petroleum is being formed. Organic matter in sediments decomposes under the impact of heat. Among the end products of this decomposition are a variety of hydrocarbons, including methane. While some methane forms along with the liquid petroleum, the main creation of methane occurs at temperatures higher than those best suited for oil. The formation of oil in sediments tends to peak around 100° Celsius (the boiling point of water); the peak of methane generation occurs at 150° Celsius.[2]

The production of methane by microorganisms can be observed in landfills. When garbage is covered with an airtight blanket of earth, methane soon begins to form. The producers of this methane are organisms that can function only in an oxygen-free environment. These minute forms of life have been given the scientific name anaerobic ("no air") microorganisms. The same process that converts decaying organic garbage into methane takes place on a vast scale throughout nature. Whenever organic debris is removed from contact with air, the methane-forming microorganisms are ready to do their work. This process has great significance for the future of mankind, because it may provide the principal key to renewable energy sources.

Methane is the most stable of all hydrocarbons. It can exist unchanged at temperatures as high as 550° Celsius (1022° Fahrenheit). All other hydrocarbons, including oil, tend to be converted into methane as temperatures increase. This process takes place in oil reservoirs at deeper locations, where the temperatures of the earth are higher. In effect, under these conditions the earth acts like a giant oil refinery, which converts heavy liquid hydrocarbons into light gaseous forms, with methane the ultimate product. Another factor that contributes to the conversion from oil to methane is the geologic age of the petroleum. The longer the oil has been in existence, the more it is likely to be converted into methane.[1, 2]

The origins of methane discussed thus far would lead us to look for this gas in sediments containing organic debris. Moreover, wherever there is oil, there is also likely to be methane. In addition, methane may be found in locations with temperatures considerably higher than those suitable for oil. Finally, the older the geological sedimentary formation, the greater the possibility of finding methane. It is conceivable that some of the largest concentrations of methane may be found in sedimentary basins below 15,000 feet, where direct conversion of organic debris to methane is combined with old oil fields that have been converted into methane through the passage of time and by means of high earth temperatures.

Coal tends to generate methane as it matures. Large quantities of methane are locked up in coal seams. This methane poses a hazard to coal mining. Thus far, very little commercial use has been made of this coal-generated methane, but it may become an important source of energy in the future. This subject is discussed more fully in the chapter dealing with unconventional sources of methane.

Some scientists have presented the hypothesis that methane may be part of the earth's original makeup. Because this hypothesis focuses on the non-biological environment, it has been labeled the "abiogenic" theory of methane's origin. This hypothesis may be summarized as follows:

(1) Ever since the earth was formed, large quantities of methane have been present in the planet's interior, with most of it being trapped in the mantle (the area between the crust and the core of the earth) and some of it having migrated to the crust (approximately the top 25 miles of the earth).

(2) Other gases which are present in the earth's mantle, such as carbon dioxide and hydrogen, may combine to generate additional methane.

(3) Methane from deep in the earth travels to the crust and surface through openings created by deep faults (fractures) and by volcanic or earthquake activities.

(4) Abiogenic (i.e., non-fossil) methane has been trapped by rock, sand, and salt formations in the crust of the earth.[3]

The abiogenic hypothesis has been championed by Professor Thomas Gold of Cornell University.[4] The validity of this hypothesis remains to be proven.

From a practical point of view, the verification of the abiogenic hypothesis would add immensely to the methane potential. Moreover, it could lead to exploration for gas in areas currently not considered good prospects, such as deep faults, volcanic areas, and earthquake zones.

Even without the substantiation of abiogenic methane, the natural gas resources of the planet are enormous. The biological sources of methane should be adequate to take care of mankind's energy needs for a long time to come. Moreover, if we move forward with the development of commercial technology for converting ocean-growing plants into methane, gas energy should be available in adequate quantities for the rest of mankind's existence.

Sources:
[1]*John M. Hunt,* Petroleum Geochemistry and Geology, *W.H. Freeman and Co., San Francisco, 1979.*
[2] Ibid., *"The Future of Deep Conventional Gas," UNITAR Conference on Long-Term Energy Sources, December 1979.*
[3] *"Abiogenic Methane: A Review," Gas Energy Review, American Gas Association, May 1980.*
[4]*Thomas Gold, "Terrestial Sources of Carbon and Earthquake Outgassing,"* Journal of Petroleum Geology, *1, 3, 1979.*

Abundant Gas Resources for America's Future

The great variety and abundance of natural gas resources in the U.S.A. justify an optimistic appraisal of the future. Conventional recoverable gas in locations less than 15,000 feet below the ground has been estimated at around 1,000 trillion cubic feet, or about fifty times annual consumption figures in recent years.[1] On a preliminary basis, gas below 15,000 feet has been put at about 200 trillion cubic feet; recent discoveries indicate that this estimate may be far too conservative. Moreover, if the federal government would expedite exploration for gas on the two billion acres it controls in Alaska, the western states, and offshore, it is likely that the overall conventional resource base would be expanded significantly.

Several thousand trillion cubic feet of gas are estimated to be present in such unconventional sources as western tight sands, Devonian shale, geopressured aquifers, and coal seams. Gasification of coal, peat, and oil shale could provide enough gas to last for centuries. Renewable sources of methane are available from organic waste products and from plants. Ocean energy farms can

23

eventually produce enough plant material for conversion into methane to help the U.S.A. achieve permanent gas energy independence.

Source:

[1] *"Potential Gas Resources in the U.S. as of Year-End 1978," by George C. Grow, Jr., Chairman, Potential Gas Committee, Gas Energy Review, American Gas Association, May 1979.*

3. Do Not Underestimate Conventional Gas

The U.S.A. has plenty of conventional gas in locations less than fifteen thousand feet into the ground. Gas resources in that category have been estimated at 1,000 trillion cubic feet by the Potential Gas Committee. This amount includes 195 trillion cubic feet of proved reserves at the end of 1979.[1]

Gas pricing policies by the federal government between the 1950's and the 1970's caused a sharp reduction in drilling activity, which led to reduced proved reserves and to lessened production of gas in recent years. Many people have drawn the erroneous conclusion from this experience that the U.S.A. is running out of gas. We should not allow politically-produced problems to distort our perspective on geological realities. The shortages of gas in the 1970's were not caused by a failure of nature's bounty, but by Washington's blindness to economic realities.

The passage of the Natural Gas Policy Act of 1978 and its economic incentives have laid the foundation for a renewal of interest in this conventional resource base. The impressive results in drilling activity, reserve additions, and production achieved in 1979 and 1980 provide considerable grounds for optimism. If political interference with exploration and production is kept to a minimum, conventional gas will make an increasingly important contribution to energy supplies. This desirable objective would be greatly facilitated if the federal government would open up more of its two billion acres for gas exploration.

The federal government owns or controls 877 million acres of land, mostly in Alaska and in the western states.[2] In addition, the government holds title to 1.1 billion acres in the outer continental shelf areas adjacent to U.S. coastal waters. According to geologists, these government-owned territories contain some of the most promising gas formations in the world. However, the federal government has followed policies that have prevented exploration for gas in most of these areas, in spite of the energy crisis that jeopardizes the very existence of this country.

According to Dean William H. Dresher of the University of Arizona College of Mines, the land portion of these federal holdings alone contain an estimated 43% of the nation's undiscovered gas.[2] In addition, vast quantities of natural gas are estimated to be present in the offshore areas owned by the federal government. The following estimates were made by the U.S. Geological Survey (USGS) and the National Petroleum Council (NPC) respectively.[3]

Offshore Natural Gas (Estimated Resource)
(trillions of cubic feet)

	USGS	NPC
Gulf of Mexico	18–91	220
Pacific	2–6	15
Atlantic	5–14	60
Alaska	9–77	195
Total	34–188	490

Since 1953, when the U.S. Congress passed the Outer Continental Shelf Act, about eight million acres have been leased for commercial exploration. This area currently provides about 15% of the natural gas produced in the U.S.A.[3] It is noteworthy that the area which has been leased constitutes *less than one percent* of the total offshore territory under federal control.

How is the government managing these vast areas in relation to gas resources? The bulk of the land holdings, some 500 million acres, have been withdrawn from exploration activities. Another 102 million acres in Alaska will be placed off limits, if recent legislation goes into effect. Similarly, large portions of the offshore areas are being withheld from gas development.[2, 3]

Government officials will point out that these areas have been set aside for parks, recreation, and wilderness. Drilling for gas in selected areas is not going to affect those uses. Once the drilling and production are completed, the area will quickly revert to its preexisting state.

Another common ploy used by the government is to claim that these areas have very little gas potential.[4] Many independent geologists would disagree with this assessment. Furthermore, there is no reliable way of knowing what the gas resources may be until exploratory drilling has taken place.

The management of public lands in relation to gas resources is too important a matter to leave in the hands of bureaucrats with limited perspectives. The American people have the right to demand more realistic behavior from their government in relation to these potential energy resources that really belong to all of us.

To sum up, geologically the U.S.A. is blessed with plentiful conventional gas resources. Political interferences with realistic gas pricing and with resource utilization have been the main obstacles to achieving increased production. As these man-made problems are resolved, the future for conventional gas looks increasingly bright.

Sources:
[1]*"Summary of United States Natural Gas Statistics for the Period 1945 to 1979," Supply & Production Supplement, Gas Supply Committee, American Gas Association, May 1980.*
[2]*"Land Use," a publication by Cities Service Company, 1979.*
[3]*"Outer Continental Shelf," a publication by Cities Service Company, 1979.*
[4]*"Preliminary Petroleum Resource Appraisal of the William O. Douglas Arctic Wildlife Range in Alaska," U.S. Department of the Interior, July 1980.*

4. Deep Gas, a Frontier with Great Promise

Methane, being the most stable of all hydrocarbons, can exist at temperatures much higher than those suitable for oil. Because of increased temperatures at greater depths, wells drilled to 15,000 feet or more have an increased likelihood of finding gas rather than oil. Major discoveries of this type have been made in the Anadarko Basin in Oklahoma and Texas, the Overthrust Belt in the Rocky Mountains, and the Tuscaloosa Trend in southern Louisiana. The Potential Gas Committee has estimated that formations at depths of 15,000–30,000 feet may contain 199 trillion cubic feet of gas.[1]

Some independent gas producers are considerably more optimistic about the prospects for finding large quantities of gas at great depths. For example, Mr. Robert Hefner III, managing partner of the GHK Corporation, claims that the deeper layers of the Anadarko Basin alone may contain between 60 and 300 trillion cubic feet of gas.[2] Similarly, Mr. Philip F. Anschutz, president of Anschutz Corporation, estimates the gas resources of the Rocky Mountain Overthrust Belt at 200 trillion cubic feet; much of this gas is situated in deep formations.[3] Mr. L. W. Funkhouser of Standard Oil Company of California, the pioneer developer of the Tuscaloosa Trend, notes that the latter covers a band approximately thirty miles wide and two hundred miles long.[4] This formation may contain 60 trillion cubic feet of gas.[5]

Considerable support for this optimistic assessment is provided by Dr. John M. Hunt, a scientist on the staff of the Woods Hole Oceanographic Institution. Dr. Hunt explains that methane remains stable at temperatures up to 550° Celsius (1022° Fahrenheit). Before such high temperatures are reached, other types of hydrocarbon, including oil, will be converted into methane.[5] It follows that methane may be the only hydrocarbon found in the deepest and oldest geologic sediments. The amounts involved could be tremendous. Dr. Hunt notes that these deep gas fields may be partly the result of giant oil fields that moved to lower depths during geological upheavals. It is noteworthy that a 60 trillion cubic foot gas resource equals ten billion barrels of oil, which would fit anybody's definition of a giant oil field. Dr. Hunt points out that the Tuscaloosa Trend may well be characteristic of other geologic formations where former oil fields have been buried at great depths.[5, 6]

Drilling wells to a depth of more than 15,000 feet is very expensive. Mr. Funkhouser notes that development wells in the Tuscaloosa Trend range in cost from $6 million to $15 million each.[4] These figures compare with $210,000 average cost for shallow wells drilled in 1978.[7] The Natural Gas Policy Act of 1978 took cognizance of these realities and deregulated gas produced from below 15,000 feet.

Deep drilling involves many technological problems. The industries serving the drilling companies have done a magnificent job in overcoming the obstacles involved. One of the most significant breakthroughs was the development of a new type of proppant by Exxon Corporation.[8] The openings (fractures) in gas fields at deep locations have a tendency to close up, thereby reducing and ultimately stopping the flow of gas. If such an occurrence were to take place at a shallow location, coarse-grained sand would be used as a proppant to keep the fracture open. However, at great depth sand is crushed to a fine powder by the pressures. After much experimentation, Exxon developed a new propping material from sintered bauxite. This proppant keeps the fractures open and the gas flowing under the greatest pressure conditions that have yet been encountered. In many instances this proppant has helped to increase production from deep wells fourfold.[2] Incidentally, the 1980 sales of this proppant are running at triple the rate of 1979, indicating sharply increased drilling activities at great depths.

While the drilling costs are high and technological challenges are extraordinary, deep wells have a major redeeming feature: their production rates are often prodigious. Wells producing 20 million cubic feet of gas a day are not uncommon. A recent report in the *Oil and Gas Journal* cites one well in the Seger (Oklahoma) field of the Anadarko Basin that was flowing 55 million cubic feet a day.[9] At 15,000 feet or more, gas is under great pressure and will readily move to the surface as long as an opening is available. Because of the great volume, the economics of deep gas production can be quite attractive.

It is noteworthy that, if a new area containing ten billion barrels of oil were discovered, all the mass media would give the event feature treatment. In contrast, major gas discoveries have aroused very little comment. How many readers of this book had previously heard of the enormous deep gas reservoirs in the Tuscaloosa Trend, the Anadarko Basin, or the Rocky Mountain Overthrust Belt? Objectively viewed, each of these areas may contain the equivalent of ten billion barrels of oil or more in the form of gas. The mass media would be well advised to get acquainted with gas and its importance to the national economy. The time has come to stop treating gas as if it were less important than oil.

Considerable leasing and exploration activities for deep gas are currently underway in the eastern part of the U.S. The area stretching from New York State to Alabama, encompassing most of the Appalachian region as well as the Blue Ridge and Piedmont ranges, has geological features favorable to finding gas. This promising region has been called the Eastern Overthrust Belt.* Among the companies involved in exploring this area are Exxon, Amoco, Mobil, Transco, Chevron, Mitchell Energy, Shell, Consolidated Gas, Columbia Gas, Atlantic Richfield, Gulf, Gaddy Oil, Allied Chemical, Phillips Petroleum, and others.[10] About ten million acres have recently been leased for oil and gas exploration in New York, Pennsylvania, Maryland, West Virginia, Virginia, Kentucky, Tennessee, and Alabama.[11]

A number of significant gas discoveries have already taken place in this region. One of the most prolific gas flows ever experienced occurred from a well drilled by Columbia Gas in Mineral County, West Virginia. This well flowed 88 million

---

*In non-technical terms, the formation of an overthrust belt may be described as follows: Movements in the deeper layers of the earth's crust cause collisions between subterranean rock formations. As a result, vast amounts of geologically old, hard rocks are thrust upward. In the process, they cover geologically young, soft formations that may contain natural gas and other hydrocarbons. Because of these geological upheavals, the gas that would normally be found in shallow locations may now be situated several thousand feet below the surface. Most of the knowledge about finding hydrocarbons in overthrust belts has been acquired in the past few years.

cubic feet of gas a day. Other gas discoveries have been made in Pennsylvania and New York.[12]

It is noteworthy that the Eastern Overthrust Belt encompasses areas that have produced gas from relatively shallow formations for a long period of time. In fact, Fredonia, New York, which gave birth to the commercial use of natural gas in 1821, is situated in the Eastern Overthrust Belt. What differentiates contemporary activities from earlier ones is the focus on deep geological formations. Refined geophysical methods for investigating gas potential deep in the earth have made a major contribution to the feasibility of finding and producing gas from this new frontier. Among the geologists most knowledgeable about the Eastern Overthrust Belt are Leonard Harris and Kenneth Bayer of the United States Geological Survey[13] and Porter Brown, chief geologist of Columbia Gas.

Deep gas has the potential of becoming one of our most important new gas frontiers. Dr. Hunt of the Woods Hole Oceanographic Institution makes the following assessment: "Deep conventional gas accumulations may well be our major fossil fuel . . . when the oil begins to run out sometime in the next century."[5] In fact, significant production from this source in various locations is imminent and is likely to increase sharply over the next several years.

Sources:
[1] *"Potential Gas Resources in the U.S. as of Year-End 1978,"* by George C. Grow, Jr., Chairman, Potential Gas Committee, Gas Energy Review, American Gas Association, May 1979.
[2] *"Deep Drilling in the Anadarko Basin,"* by Robert A. Hefner III, Petroleum Engineer International, *March 1979.*
[3] *"A Review of the Overthrust Belt,"* by Philip F. Anschutz, *address to the American Gas Association, October 22, 1979.*
[4] *"The Deep Tuscaloosa Gas Trend of South Louisiana,"* by L.W. Funkhouser, F.X. Bland, and C.C. Humphris, Jr., *paper presented at*

the national meeting of the American Association of Petroleum Geologists, June 9, 1980.

[5]*"The Future of Deep Conventional Gas,"* by John M. Hunt, UNITAR Conference on Long-Term Energy Sources, December 1979.

[6]Petroleum Geochemistry and Geology, *by John M. Hunt, published by W.H. Freeman and Company, San Francisco, 1979.*

[7]*"Primer on Natural Gas and Methane,"* Scientists' Institute for Public Information, November 15, 1979.

[8]*"Use of High-Strength Proppant for Fracturing Deep Wells,"* by Claude E. Cooke, Jr., John L. Gidley, and Dean H. Mutti, Exxon Company. This paper was presented at the 1977 Deep Drilling and Production Symposium of the Society of Petroleum Engineers held in Amarillo, Texas, April 17–19, 1977.

"High-Strength Proppant Extends Deep Well Fracturing Capabilities," by C.E. Cooke, Jr. and J.L. Gidley, Exxon Company U.S.A.

[9]*"Deep Drilling Surge Hits Anadarko Basin,"* by John C. McCaslin, Oil & Gas Journal, August 4, 1980.

[10]*"New Gas Potential for Appalachia,"* Gas Energy Review, American Gas Association, December 1979.

[11]*"An Overthrust Belt Explored in the East,"* by Richard D. Lyons, New York Times, March 24, 1980.

[12]*"New Finds Heat Appalachian Basin Interest,"* Oil & Gas Journal, February 11, 1980.

[13]*"Broader Expanse Seen for Eastern Overthrust,"* Oil & Gas Journal, October 15, 1979.

5. Supplemental Sources of Conventional Gas

During the past decade the gas industry has experienced supply shortages. While the outlook for more plentiful supplies of U.S. natural gas has improved since the passage of the Natural Gas Policy Act of 1978, the industry is actively seeking additional amounts of gas from supplemental sources. The latter include gas from Alaska, imports from Canada and Mexico, and liquefied natural gas (LNG) from countries with gas surpluses.

Gas resources of Alaska have been estimated at 221 trillion cubic feet, of which 31.6 trillion cubic feet are proved reserves.[1] Two projects have been planned to bring some of this gas to the continental U.S.A. One involves the transportation of liquefied natural gas (LNG) from southern Alaska to California. The other requires the construction of a gas pipeline from northern Alaska through Canada to the lower 48 states.

A liquefaction plant in Nikiski, Alaska will liquefy natural gas from the Cook Inlet area. From Nikiski the gas will be shipped by tanker to Point Conception, California, where it will be regasified and put into existing pipelines. It is expected that this source will contribute about 150 billion cubic feet of gas by 1990 and some 600 billion cubic feet by the year 2000.[2]

A 4,800 mile pipeline is planned to bring gas from the Alaskan North Slope to the lower 48 states, with one branch going to the Midwest and another to California. This project is estimated to cost around $23 billion. Construction will take about three years. This pipeline could supply 1.4 trillion cubic feet of gas by 1990. A second pipeline in the same general area could increase capacity to three trillion cubic feet by the year 2000.[3]

Canada and Mexico have substantial amounts of natural gas available for export. The gas from these sources can be connected with existing U.S. pipelines, which have sufficient capacity to carry the loads involved. As a result, capital investments are minimal. These imports are particularly useful because of their geographic proximity and their relative immunity from adverse external developments. Moreover, both Canada and Mexico are good customers for U.S. products and services, including those involved in the production of natural gas.

Gas imports from Canada have averaged around one trillion cubic feet a year since 1971.[4] The Alaskan pipeline project, which will cross over Canadian territory, will open up significant additional Canadian gas resources for possible exports. From all available indications, Canada has huge potential gas supplies.[5, 6]

Mexico is currently exporting 300 million cubic feet of gas per day to the U.S.A. These shipments were initiated in January 1980.[7] The gas resources of Mexico are very large and could supply additional amounts to the U.S.A.[8]

On a worldwide basis, there is currently a large oversupply of natural gas. In a number of oil-producing countries, several trillion cubic feet of gas are flared each year, a waste of a valuable resource. In other places, vast amounts of gas are shut in because of lack of markets. Liquefying the gas and transporting it by tanker to markets that want it is a rational approach to energy utilization that benefits both producers and consumers. The main constraints to large-scale commerce in liquefied natural gas include the high initial capital costs and governmental restrictions.

Because of sizeable North American gas resources, the U.S. is not dependent on liquefied natural gas imports from other parts of the world. However, under appropriate economic conditions, the U.S. could utilize as much as two trillion cubic feet per annum of liquefied natural gas by the year 1990. This amount could be doubled by 2000.[1]

Sources:

[1]*"The Gas Energy Supply Outlook: 1980–2000," A Report of the A.G.A. Gas Supply Committee, October 1980.*

[2]*"Alaskan Gas," Public Information Division, American Gas Association, February 1980.*

[3]*"Alaskan Gas Pipeline Enters Financing Stage," Financial Quarterly Review, A.G.A., October 1979.*

[4]Monthly Energy Review, *July 1980,* U.S. Department of Energy, Washington, D.C.

[5]*John A. Masters, "Deep Basin Gas Trap, Western Canada,"* Bulletin of the American Association of Petroleum Geologists, *August 1978.*

[6]Canadian Natural Gas Supply and Requirements, *National Energy Board, Minister of Supply Services, Canada, February 1979.*

[7]*"Border Gas Consortium Receives First Supplies of Mexican Gas," American Gas Association Monthly, June 1980.*

[8]*Minister Florencio Acosta, "The Role of Oil in the Mexican Development Plans," Embajada de Mexico, June 12, 1979.*

6. Large Treasures of Unconventional Gas Await Development

The U.S.A. is well endowed with such unconventional sources of gas as western tight sands, Devonian shale, methane in coal seams, and geopressured aquifers (water-bearing rocks). The potential resources involved are enormous. Many technological and other problems remain to be solved to convert these resources into major contributors to U.S. gas supplies. The gas industry expects to spend some $30 billion over the next twenty years to develop these resources. If these programs achieve their objectives, unconventional sources might contribute 5 trillion cubic feet of gas by the year 2,000.[1] Significant breakthroughs in technology could speed up developments and lead to impressive results even sooner.

Western Tight Sands

The Rocky Mountain regions of Wyoming, Colorado, and Utah contain large amounts of gas trapped in tight sands. Estimates of this resource range from 400–600 trillion cubic feet.[2] The tight sandstone formations surrounding the gas have to be fractured to make possible economic recovery of this resource. Research is being done to improve fracturing techniques. Among the participants in this research are Gas Research Institute, Mobil Oil Corporation, Mitchell Energy Corporation, and Petroleum Technology Corporation.[2]

To achieve commercial production of gas from tight sands, massive amounts of fluid and propping materials are injected under pressure into the formation. This technique fractures the sandstones and helps to release some of the gas. Fracturing may have to be done several times to achieve satisfactory results. Practical know-how of each specific gas-containing formation is essential if optimum recovery is to be achieved.

Estimates by the American Gas Association Gas Supply Committee indicate that a modest amount of production from this source may start by 1984.[3] With advanced technology and attractive market prices ($4.50 per thousand cubic feet in 1979 dollars), output could reach 1 trillion cubic feet per annum in 1990 and 5 trillion cubic feet per annum in 2000. It should be noted that the figures cited here are not the same as those used for calculating capital expenditures. (See chapter 23) The latter are based on more conservative assumptions. If technological and economic developments are favorable, capital expenditures would have to be raised to achieve the 5 trillion cubic feet level.

Devonian Shale

A large part of the Appalachian Basin is underlaid with brown shales which contain significant amounts of gas. Recent estimates by experts place the recoverable gas in the range of 60–600 trillion cubic feet.[3]

Gas contained in Devonian shale flows slowly and makes economic recovery difficult. To speed up the flow, a variety of fracturing techniques are employed, including explosives and massive hydraulic fracturing. The latter involves injecting fluids and sand under great pressure into the shale. These procedures of course add to production costs. The Natural Gas Policy Act of 1978 gave recognition to these realities by lifting price controls from newly developed Devonian shale gas.

The Morgantown Energy Research Center of the U.S. Department of Energy is engaged in research programs to facilitate greater production of gas from Devonian shale. The Appalachian Regional Commission has supported shale development. Private firms active in research on Devonian shale include Columbia Gas System, Mitchell Energy Corporation, Consolidated Gas Supply Corporation, and Thurlow Weed & Associates, Inc.[2]

It should be noted that the Appalachian Basin has produced gas for the past 120 years. Gas production averages about 400 billion cubic feet a year. This production comes from thousands of wells, which average 79,000 cubic feet of gas a day, a very low rate of production by gas industry standards. In fact, most wells in the Appalachian Basin qualify as "stripper wells," because they produce less than 60,000 cubic feet of gas a day.[4]

Assuming advanced technology and a market price of $4.50 per thousand cubic feet (in 1979 dollars), Devonian shale is estimated to contribute 1 trillion cubic feet of gas per annum by the year 2,000. While the annual production rates from Devonian shale may appear modest, such gas resources have the virtue of being productive for a very long time. Devonian shale may contribute significantly to U.S. gas production throughout the 21st century.[3, 5, 6, 7]

The figures cited in the preceding paragraph do *not* include current production from Devonian shale.

Methane in Coal Seams

Large quantities of methane are trapped in coal seams. This resource base has been estimated at 500 trillion cubic feet of gas.[3] Recoverable resources are estimated at 10–60 trillion cubic feet, the larger amounts being contingent on the development of advanced technology and the availability of high market prices. It has been estimated that by 1990 approximately 200 billion cubic feet per annum of methane may be recovered from coal seams; under favorable economic and technological conditions, the production rate could exceed one trillion cubic feet per annum by the year 2000.[3]

Methane must be removed from coal seams to make safe mining possible. Hitherto, methane has been vented into the air, wasting this valuable resource. In recent years the U.S. Bureau of Mines and the Department of Energy have done research on techniques for recovering this methane. The technological problems have been largely solved.[8]

Methane recovery from coal beds tends to be slow and may interfere with the most economic rate of coal production. This problem can probably be solved with improved technology and by appropriate scheduling of methane recovery prior to the start of mining.

The biggest obstacle to the speedy utilization of methane from coal seams concerns the legal status of this resource. In many states there is confusion about ownership rights. Do the owners of the coal mines also own the methane, or does methane come under the provisions of oil and gas mineral rights? The early resolution of this issue would facilitate making this valuable resource available to help meet U.S. energy requirements.[2, 9]

Geopressured Aquifers

The northern Gulf of Mexico coastal region contains large amounts of gas dissolved in hot brine in geopressured zones. Sedimentary deposits in this area range to 50,000 feet in thickness. The geopressured zones are 5,000 to 15,000 feet below the surface. Estimates of the amounts of methane contained in geopressured aquifers range from 860 to 100,000 trillion cubic feet. There is also a wide spectrum of views as to the amounts that may be economically recovered; estimates run from 42 to 2,000 trillion cubic feet.[3]

Many technical problems remain to be solved before gas can be economically recovered from geopressured formations. By the year 2000, annual production from this source will probably range between 20 and 200 billion cubic feet. Under optimistic assumptions, production could reach 1 trillion cubic feet.[3]

The Department of Energy, Institute of Gas Technology, University of Texas at Austin, and Louisiana State University are engaged in research on this resource. The following topics are being investigated:

(1) Assessment of geothermal-geopressured resources in Texas and Louisiana;

(2) Testing the feasibility of production from new wells in Brazoria County, Texas and Cameron Parish, Louisiana; and

(3) Laboratory research to determine reservoir characteristics and to make forecasts of the economic and production profiles of geopressured wells.[2]

Because of the huge resource potential, geopressured gas has aroused considerable scientific interest. Several papers presented at the UNITAR Conference on Energy and the Future, in Laxenburg, Austria, July 1976, dealt with that topic.[10, 11] The American Gas Association devoted its October 1978 Gas Supply Review to geopressured gas.[12] *Grid,* the publication of the Gas Research Institute, contained a detailed article entitled "Natural Gas From Geopressured Zones," in the March 1979 issue.[13]

Hydrates

In addition to the preceding unconventional sources of methane, mention should also be made of gas hydrates, which are solid, ice-like compounds in which methane is entrapped and bound to water molecules. Permafrost zones and cold ocean sediments are the likely places where gas hydrates may be found. Dr. Hunt mentions gas hydrates in his book.[14] The National Science Foundation and the Office of Naval Research have been doing research on the subject. Scientists of the Soviet Union are probably most knowledgeable in this area. The resource estimates are very large, but no proven production technology has been developed in the western world.[2] Considerable information about the subject is presented in the August 1979 Gas Energy Review.[15]

Sources:
[1] *"A Forecast of Capital Requirements of the U.S. Gas Utility Industry to the Year 2000," Policy Evaluation & Analysis Group, American Gas Association, April 20, 1979.*
[2] *"New Technologies for Gas Energy Supply and Efficient Use," American Gas Association, April 1979.*
[3] *"The Gas Energy Supply Outlook: 1980–2000," A Report of the A.G.A. Gas Supply Committee, October 1980.*
A good basic analysis of western tight sands is presented by Richard F. Meyer of the U.S. Geological Survey in a chapter entitled "The Resource Potential of Gas in Tight Formations," which was published in the book The Future Supply of Nature-made Petroleum and Gas, *Pergamon Press, Elmsford, New York, 1977.*
[4] *Porter Brown, "Oil and Gas Potential of the Appalachian Thrust Belt," presented to the Potential Gas Committee, Scottsdale, Arizona, October 19, 1979.*

[5] *"Gas from Devonian Shale—Status and Outlook,"* Gas Supply Review, *American Gas Association, May 1978.*

[6] *"Gas from Shales Rich in Organic Detritus,"* by Wallace DeWitt, Jr., *UNITAR Conference on Long-Term Energy Sources, Montreal, December 1979.*

[7] *"Chemical and Bacterial Treatment of Devonian Shale for Gas Recovery and Production,"* by Teh Fu Yen, ibid.

[8] *"Natural Gas from Caolbeds,"* by Troyt B. York, ibid.

[9] *"Methane from Degasification of Coalbeds,"* Gas Supply Review, *American Gas Association, June 1977.*

[10] *"The Supply of Natural Gas from Geopressured Zones: Engineering and Costs,"* by Myron H. Dorfman, The Future Supply of Nature-Made Petroleum and Gas, *Pergamon Press, Elmsford, New York, 1977.*

[11] *"Gas in Geopressured Zones,"* by Paul H. Jones, ibid.

[12] *"Special Geopressured Gas Issue,"* Gas Supply Review, American Gas *Association, October 1978.*

[13] *"Natural Gas from Geopressured Zones,"* by Robert B. Rosenberg *and John C. Sharer, Gas Research Institute,* Grid, *March 1979.*

[14] Petroleum Geochemistry and Geology, *by John M. Hunt, published by W.H. Freeman and Company, San Francisco, 1979.*

[15] *"Gas from Natural Gas Hydrates,"* Gas Energy Review, American *Gas Association, August 1979.*

7. Gasification of Coal and Peat

The gas industry must plan its supply programs many years in advance. Such planning is facilitated if gas supplies have the following characteristics:

(1) plentiful raw materials;

(2) location near existing markets or pipelines;

(3) availability of proven technology;

(4) potential for expansion;

(5) competitive costs; and

(6) secure control over production facilities. All of these characteristics apply to gas derived from coal and peat.

Coal is the most plentiful fossil fuel in the United States. The amount of coal that could be economically converted into gas is so large that there would be no foreseeable raw material supply problem. Gasification processes do not require high-grade coal. For example, the first modern commercial coal gasification project in the U.S. uses lignite, which would be economically unacceptable for most other applications. Many of the ingredients in coal that would normally cause pollution are removed during the gasification process and become valuable byproducts, such as sulfur and anhydrous ammonia. Gasification procedures can also be applied to underground coal seams that cannot be mined. Much research is being done on such in situ gasification, which takes place in the location where the coal is situated. Depending upon what categories of coal are included for potential gasification use, the estimates range from 1,000 to 10,000 trillion cubic feet of gas from that source.[1]

Peat ranks second only to coal as a material suitable for gasification. Peat covers some 52 million acres in the U.S. This acreage contains some 120 billion tons of peat, which could be converted into 1,440 trillion cubic feet of gas. About half of the peat is located in Alaska. Much of the rest is situated in states that are not generally well endowed with hydrocarbon resources, including Minnesota, Michigan, Wisconsin, Maine, New York, and Massachusetts. The development of peat gasification should be encouraged to help bring about a more balanced energy resource base for the northern and northeastern states.[2]

Much of the coal and peat in the continental U.S. is located near existing gas pipelines or near major markets. Transportation of the gas from the gasification facilities to the end users would not present any insurmountable problems.

Technologies for converting coal and peat into gas have been proven both in the U.S. and in other countries. Private companies in the U.S. have spent over $130 million for research on coal gasification.[3] Additional amounts have been spent by government agencies.

Much of the pioneering work on peat gasification has been carried on by the Minnesota Gas Company, together with the Institute of Gas Technology (IGT).[4] The economics of peat gasification look promising. It has been found that peat is easier

46

to gasify than coal.[5] IGT discovered that the carbon in peat converts more readily into hydrocarbon gases than does coal. Moreover, only low hydrogen pressures are required to convert peat to gas. Both of these factors contribute to production efficiency and result in favorable cost comparisons.[6]

Once commercial facilities for coal and peat gasification are in operation, the potential for expansion is vast. The only practical limitations that exist concern the economic costs and their comparisons with costs of gas from other sources. The likelihood of technological advances and cost reductions for coal and peat gasification is noteworthy. In fact, second generation coal gasification equipment is already being developed. While gasification is unlikely to be as economical as the recovery of natural gas from large fields, the existence of gasification facilities may place a limit on the prices the U.S. has to pay for imported oil.

The experience of the past several years has dramatized the importance of long-range gas supplies. Realistic business judgment would indicate the advisability of gas pipeline and utility companies owning part of the gas they are handling. Coal and/or peat gasification plants can fulfill this function admirably. Apparently many gas companies have reached this conclusion. By the year 2000, up to 40 coal gasification projects, involving the expenditure of almost $60 billion, may be built. Most of the early projects will be joint ventures by several utilities.[7]

The first commercial coal gasification facility is currently being constructed in North Dakota by the Great Plains Coal Gasification Association, a joint venture of American Natural Resources Company, Tenneco, Peoples Energy, Columbia Transmission Corporation, and Transco. Assuming no unforeseen delays, this facility will go on stream in 1984 and will produce 125 million cubic feet of gas a day. Total cost of the project is estimated at $1.5 billion. The plant will use 22,000 tons of lignite coal a day. This low-grade coal will be converted into pipeline quality gas using the Lurgi process, which was originally developed in Germany. Several byproducts are recovered, including sulfur, carbon dioxide, and anhydrous ammonia. Environmental pollution is minimized with this procedure. The plant is designed for a potential doubling of gas output.[8]

It is noteworthy that the President's Council on Environmental Quality has published data showing that commercial scale coal gasification plants will emit only one-tenth the air pollutants of equivalent coal-fired electric power plants, even if the latter use the best available pollution control technology.[9] Moreover, such coal gasification facilities would use considerably less water than coal-fired electric generating plants. The water use has been estimated at one-tenth to one-fifth of the amounts necessary for regular electric generation.[9]

The first peat gasification plant capable of producing 80 million cubic feet of gas a day is expected to go on stream in 1985. If a maximum effort is made to utilize peat, this resource could contribute 900 billion cubic feet of gas per annum by the year 2000.[10]

Coal gasification is expected to contribute increasing amounts of gas starting in the mid-1980's and may provide over three trillion cubic feet annually by the year 2000.[10]

Ownership of coal and peat gasification plants will give gas companies a significant degree of control over their own destinies.

Sources:
[1]*"The Future for Gas Energy in the United States," American Gas Association, June 1979.*
[2]*"Peat: A Major Energy Resource to Meet U.S. Clean Fuels Needs," Institute of Gas Technology, August 1979.*

[3] *"Fact Book: Synthetic Pipeline Gas from Coal,"* American Gas Association, September 1979.

[4] *"New Technologies for Gas Energy Supply and Efficient Use,"* American Gas Association, April 1979.

[5] *"Synthetic Natural Gas from Peat,"* by Arnold M. Rader, Gas Supply Review, American Gas Association, December 1977.

[6] *"Peat and the Environment,"* Institute of Gas Technology, January 1979.

[7] *"The Forecast of Capital Requirements of the U.S. Gas Utility Industry by the Year 2000,"* American Gas Association, April 20, 1979.

[8] *"The Great Plains Story,"* Venture, published by American Natural Resources System, Summer 1980.

[9] *"A Western Regional Development Study: Primary Impacts,"* prepared for the Council on Environmental Quality by Radian Corporation, August 1975.

[10] *"The Gas Energy Supply Outlook: 1980–2000,"* A Report of the A.G.A. Gas Supply Committee, October 1980.

8. Methane from Garbage, Sewage, and Animal Wastes

Plants have the unique ability to utilize the sun's energy for their own growth. This transformation of solar energy into plant life is brought about through the process of photosynthesis. When plants have completed their life cycle, the residue can be transformed into methane by specialized microorganisms that function in an oxygen-free environment. In this manner, garbage, sewage, and animal waste can become sources of gas energy.

In nature, microorganisms break down large amount of organic debris into simple molecules, including methane. This process reduces the volume of waste products. Without the action of these microorganisms, the earth might long ago have been buried in its own organic debris. In recent years, human beings have begun to emulate nature's waste recycling procedures for the production of methane.

Thermal decomposition (pyrolysis) of organic debris into methane is an alternative to microbiological methane generation. Economic and environmental factors have to be evaluated in each case to determine which procedure is more advantageous. The balance of this chapter will focus on the microbiological approach.

Gas from Garbage

The recovery of methane from garbage entails significant benefits. This procedure provides reliable energy in the vicinity of major consuming markets. The removal of methane from garbage sites reduces environmental hazards, improves the quality of life for the surrounding communities, and generates income to local governments.

In the U.S. garbage is being generated at the rate of 3.5 pounds per person each day. In 1977 the Environmental Protection Agency estimated total garbage generation at 135 million tons a year. The amount is expected to increase to 225 million tons by 1990.[1]

Not all of this garbage can be converted into methane. About one-fourth consists of inorganic materials, such as metals and glass, which make no contribution to methane production. Only organic garbage that is situated in covered landfills of adequate dimensions will generate methane of economical value. By 1990, it is estimated that between 50 and 300 billion cubic feet of gas per annum could be obtained from garbage.[2, 3] If the more optimistic estimates were to be realized, this methane could supply about 9% of the gas demands in the 65 largest cities of the United States.[3]

Gas from garbage is only partially methane; the other major component is carbon dioxide. Generally, untreated gas from garbage contains 50–56% methane and 40–45% carbon dioxide. Small amounts of other trace elements are also present.[4]

If an industrial user or utility is situated near the landfill site, the gas can be channeled with very little further treatment to such a customer. However, if the gas is to be added to the regular pipeline network, it has to be upgraded to meet pipeline specifications. This procedure involves the removal of the carbon dioxide and any trace elements that are not allowed in pipeline gas. The technology poses no problems, but the procedures add to the costs.

California deserves credit for pioneering the commercial development of methane recovery from garbage. The first operational facility was constructed in Palos Verdes, California, in 1975. Since that time, a number of other localities in California have developed similar projects.[2] New York City and Chicago will soon join their ranks.

New York City expects to have a methane facility in operation by the spring of 1982. The methane will be recovered from a landfill on Staten Island; the Brooklyn Union Gas Company will purchase the gas. Once the operation is in full swing, New York City expects to collect $1 million a year in royalties.[5] Other benefits to the city include improved environmental conditions and a more effective way of coping with the huge garbage load.

In addition to its economic values, a methane recovery facility helps to reduce air pollution and other environmental hazards. An untreated landfill will release noxious fumes into the air. Moreover, if the methane is not collected, it will seep into the atmosphere and may pose a fire hazard. People who are concerned about a wholesome environment should familiarize themselves with the helpful role methane recovery facilities can play in their communities.

Gas from Sewage

Sewage, which consists of about 99% water and 1% organic materials, can be used to grow water hyacinths, algae, or other aquatic plants, which in turn can be transformed into methane by microorganisms. If enough sewage were treated in this fashion in the U.S.A., an estimated 300 to 500 billion cubic feet of methane could be produced annually.[6, 7] This approach has the additional advantage of helping to clean up the environment and improving the management of waste treatment.

Methane from Animal Wastes

Concentrated amounts of cattle manure, such as those found in large feedlots, provide a good source of methane. Thirty-five million tons of such manure are produced in the U.S.A. annually. It is estimated that this manure could be converted into 200 to 300 billion cubic feet per annum of methane. Some commercial production of methane from manure has been started; more is expected in later years.[7]

Animal manures and other agricultural waste products can be used to produce methane as well as fertilizer.[8] A number of foreign countries are utilizing such procedures and are engaged in active research programs to facilitate use of farm products for methane generation.[9, 10, 11]

Other biomass sources of methane may be developed, depending primarily upon economic considerations. Whenever methane can be produced along with other end products, the economics of biomass conversion may become attractive. The technology for biodigestion of organic waste materials is well known. It is likely that more applications of this approach will be found in the future as more people become aware of the possibilities.

Sources:
[1] *"Resource Recovery and Waste Reduction," Fourth Report to Congress, Environmental Protection Agency, 1977.*

[2] *"Methane from Landfills,"* Gas Energy Review, American Gas Association, April 1980.

[3] *"Primer on Natural Gas and Methane,"* Scientists' Institute for Public Information, November 15, 1979.

[4] *"Upgrading Landfill Gas to Pipeline Specifications,"* by Robert H. Collins, III, paper presented to the American Society of Mechanical Engineers, Houston, Texas November 6, 1978.

[5] *"Methane Gas Recovery Projects,"* The City of New York, Department of Sanitation, September 18, 1979.

Other sources on gas from landfills worth noting:

"Recovering Gas from Landfills: Resource Potential and Institutional Barriers," by Philip J. Mause, James A. Hayes, Richard T. Williams, and Norman A. Pedersen, of the firm Kadison, Pfaelzer, Woodard, Quinn & Rossi, Washington, D.C., March 1980.

"Methane Gas Recovery from Landfills—A Worldwide Perspective," by Robert A. Colonna, which was published in the book The Future Supply of Nature-made Petroleum and Gas, *Pergamon Press, Elmsford, New York, 1977.*

"Methane Gas Recovery, Processing and Utilization from Sanitary Landfills in the United States," by Kenneth K. Hekimian and Stanley K. Katten, UNITAR Conference on Long-Term Energy Sources, Montreal, December 1979.

"Urban and Industrial Waste—Energy Production Potential," by John T. Pfeffer, ibid.

[6] *"Supplemental Gas from Biomass,"* draft of a document by the American Gas Association.

[7] *"The Gas Energy Supply Outlook: 1980–2000,"* A Report of the A.G.A. Gas Supply Committee, October 1980.

[8] *"Methane from Manure,"* by T.H. Hutchinson, UNITAR Conference on Long-Term Energy Sources, Montreal, December 1979.

[9] *"The Energy Plantation and the Photosynthesis Energy Factory: Multifunctional Bio-Energy Systems,"* by Malcolm D. Fraser, Jean F. Nenry, Louis C. Borghi, and Norman J. Barbera, ibid.

[10] *"Bio-Fuels Projects at the Royal Institute of Technology of Stockholm, Sweden,"* by Pehr Bjornbom, Stefan Engstrom, and Olle Lindstrom, ibid.

[11] *"Potential of Anaerobic Digestion in Biogas Production in Developing Countries,"* by Sermpol Ratasuk, ibid.

9. The Production of Methane from Ocean Plants

Giant kelp and other marine plants may provide the ultimate solution to the energy problem. It has been estimated that 55,000 square miles of kelp farms could provide the raw materials for producing 20 trillion cubic feet of methane each year, an amount equal to total current U.S. gas consumption.[1] Considerable research is being devoted to developing suitable procedures and technology. Scientific progress has already brought encouraging results.

About 4.5 miles offshore from Laguna Beach, California, an experimental test farm has been constructed that may hold the key to almost limitless amounts of methane in the future. On this site, giant brown kelp plants *(Macrocystis pyrifera)* are growing and reproducing normally in their transplanted setting. The term "giant" is appropriate for this kelp plant. It is the largest of the marine algae, attaining a length of 200 feet in the adult state. The plant grows as much as two feet in a single day. These rapid growth characteristics indicate the high efficiency of the plant in utilizing solar energy and nutrients from the ocean environment. Scientists have found that this kelp absorbs as much as 99 percent of the incident sunlight through several layers of blades (leaves) which float near the surface and form a canopy. The kelp is easily harvested by removing the top ten feet of the plant. The removed portions are soon replaced by new fronds growing from the base of the plant. Reproduction is by means of spores. Once a kelp farm has been established, it is likely to be self-sustaining indefinitely, with no new plantings required.[2]

The harvested kelp is placed in large airtight containers, called biodigesters. Microorganisms ingest the kelp and convert it into methane. It has been found that anaerobic bacteria thrive on kelp in sea water. According to scientists involved in the project, the methane yield is the highest of any biomass material.[3] Marine kelp have a major advantage over terrestial plants, which require lignin cellulose for structural support. Lignin is not suitable for conversion into methane by anaerobic microorganisms; kelp is free from this substance.[4]

Research on this marine farm is being sponsored by the Gas Research Institute, with research being done by scientists supplied by General Electric, California Institute of Technology, the U.S. Department of Agriculture, and the Department of Energy. The engineering work involved in constructing the farm and making it seaworthy under ocean conditions is being conducted by Global Marine Development Inc. The project has many of the high-technology characteristics of the space program. In fact, the main contract is held by the Re-Entry and Environmental Systems Division of General Electric.

The overall research program is divided into three phases:

(1) Concept validation (1973–1982);

(2) Process development (1982–1987);

(3) Commercial demonstration (1992).[3]

Scientists working on the program are encouraged by the

results that have been achieved. They have expressed the belief that within the next three years the feasibility of the marine energy farm concept will be determined.[3] Considerable improvements are also expected in anaerobic digestion, with particular emphasis on speeding up the process and increasing yields of methane.[5]

The positive results achieved thus far have encouraged the exploration of additional site studies for marine plant research in waters off New York State and other Atlantic areas, as well as coastal waters in the Gulf Coast, Hawaii, and Alaska.[2]

A 1,000 acre marine farm demonstration facility is expected to be constructed by 1993. This unit will help test the economic aspects of the kelp-to-methane process. Preliminary estimates on the basis of laboratory studies indicate that methane from kelp will be competitive with alternative systems of producing substitute natural gas. The cost comparisons are even more favorable if allowance is made for the other economic values that will result. The solid remains of biodigestion can be sold as animal feed supplements and as fertilizer. Moreover, commercial marine farms will attract a large fish population, which can provide significant sources of seafood.[1]

The work on marine plant sources of methane has profound implications. If successful, it can help lay the foundation for an assured supply of methane for all of mankind for the rest of time. It can banish the fears of running out of energy. It should also facilitate more realistic planning for the immediate future. If major renewable sources of methane are within reach, we can utilize our fossil fuel base with greater assurance.

Sources:
[1] *"Methane from Seaweed," by J.C. Sharer and A. Flowers,* Grid, *Gas Research Institute, January 1979.*
[2] *"Marine Biomass Energy Project," by James R. Frank and Joseph E.*

Leone, presented to the *AGA Transmission Conference, Salt Lake City, Utah, May 6, 1980.*

[3] *"A Review of the Energy From Marine Biomass Program,"* by Armond J. Bryce, presented to the *Biomass Symposium, Institute of Gas Technology, August 14–17, 1978.*

[4] *"Marine Biomass Energy Project,"* Joseph E. Leone, presented to the *Marine Technology Society, New Orleans, October 11, 1979.*

[5] *"Studies Improve Biomass to SNG Conversion,"* by James R. Frank, Hydrocarbon Processing, *April 1980.*

Other sources:

"Marine Biomass Research Program Testimony," by Dr. Ab Flowers, Director, Gas Supply Research, Gas Research Institute, to the Subcommittee on Oceanography, U.S. House of Representatives, September 26, 1979.

"Marine Macroscopic Plants as Biomass Sources," by Wheeler J. North, UNITAR Conference on Long-Term Energy Sources, Montreal, December 1979. Professor North of the California Institute of Technology (Corona del Mar) is in charge of investigating kelp growth and nutrition for the Marine Biomass Energy Project.

10. Efficient Use of Gas is Essential

The increasing costs of all types of energy, including gas, should provide strong incentives to increase the efficient operation of existing equipment, or to replace wasteful devices with energy-saving ones. The gas industry is sponsoring research to facilitate this process. A number of technological improvements have been made which will enable consumers to achieve significant savings on gas bills. Recently announced energy savers include warm air furnaces that recover heat which normally passes up the chimney plus a new burner concept called pulse combustion. Among promising future devices are gas-fired heat pumps and onsite fuel cells.

The performance of improperly installed and poorly maintained residential and commercial gas heating units can be improved through professional attention. In many instances, the elimination of such shortcomings could result in significant savings of money and energy. Before considering the purchase of new equipment, Americans would be well advised to make a thorough check of existing devices to ascertain the possibility of significant operating improvements through better maintenance. If all owners of residential and commercial buildings were to follow this procedure, it would have the same effect as adding ten percent or more to our national gas supplies. If you are seriously interested in reducing your monthly gas bills, you should take care of this matter without further delay. You may check with your local gas utility for further details and for suggestions of qualified professionals to look after your equipment.

Keeping the preceding comments in mind, we can now examine some of the new energy-saving devices that can improve efficiencies even more, though at considerable initial expense.

Pulse Combustion

Pulse combustion uses gas to generate heat in a sealed chamber. It operates similar to an internal combustion engine and uses a spark plug for ignition of the gas. This approach eliminates the need for chimney, burner, or pilot light. Most of the heat remains within the combustion chamber, which accounts for the high efficiency. The pulse combustion boiler, which is currently being marketed, is claimed to have an efficiency of 91–94%.[1] A warm-air furnace using a similar approach will be marketed in the latter part of 1981.[2] Marketing of these high-technology devices is done through certified heating contractors, who should be able to provide detailed information. The fuel savings that can be achieved with these units can range from 30–50% on fuel costs, depending on the efficiency of the equipment they replace.

The author notes that he has no personal familiarity with these devices, nor is he equipped to answer questions on technical matters. All queries should be directed to the manufacturers (see footnotes) or to certified heating contractors. The comments made here are for informational purposes only; they should not be construed as an endorsement for any commercial product.

The Heat Pump

One of several gas-fired heat pump projects currently being developed includes a heat engine, which powers a refrigeration cycle that produces heat in the winter and cooling in the summer. This type of equipment will provide substantial savings in energy consumption compared with conventional furnaces and air conditioners. The Gas Research Institute is doing research and development on gas-fired heat pumps for residential and commercial dwellings. It is hoped that they will be ready for marketing by 1985.[3]

The Fuel Cell

Fuel cells use natural gas to generate electric power at locations such as multi-family residences, commercial buildings, or industrial plants. The technology involved grew out of the space program, which utilized fuel cells for providing electricity to space vehicles. Fuel cells are highly efficient, because they avoid the energy losses involved in the generation of electricity as well as its long-distance transmission. They can also make use of heat that would otherwise be wasted. In a test made in a multi-unit apartment building in New England, a fuel cell combined with a heat pump used less gas to produce all the heating and electrical requirements than was needed by a conventional furnace to produce heat alone. Field tests are expected to start in 1981. Gas-fired fuel cells may be commercially available in the mid-1980's.[4]

Residences and commercial buildings can save gas and

money if they make use of advanced equipment and operate their present heat and power sources with optimum efficiency. Tax policies should be structured in ways to provide inducements for energy efficiency.

The gas industry is taking an active role in fostering efficient use of its product. Much of the research on fuel-saving devices has been sponsored by the gas industry. Gas utilities are eager to maintain their record of providing the lowest cost energy service to their customers. In a period when gas prices will inevitably rise, the best way to accomplish this goal is through greater efficiency at the point of end use.

Sources:
[1]*Hydrotherm, Inc., Rockland Avenue, Northvale, New Jersey 07647.*
[2]*Lennox Industries, Inc., P.O. Box 400450, Dallas, Texas 75240.*
[3]*James E. Drewry, "Gas-Fired Heat Pumps,"* Grid, *Gas Research Institute, September 1978.*
[4]*Richard T. Sperberg, "Onsite Fuel Cells,"* Grid, *Gas Research Institute, October 1979.*

11. Gas Pipelines and Storage

Over one million miles of pipelines connect the nation's gas fields with the 47.6 million customers of the gas utilities. The pipelines have sufficient capacity to carry about 23–24 trillion cubic feet of gas per year. They are backed up by storage facilities that can hold 7.3 trillion cubic feet of gas.[1]

The bulk of the nation's gas pipelines connect the gas fields in the southwestern part of the country with markets in the midwest and northeast. New pipelines are being constructed or planned from the Rocky Mountain Overthrust Belt and from Alaska to provide additional gas supplies to major consuming areas in California and the Midwest. The pipeline system in place now could transport an additional 3–4 trillion cubic feet of gas per year. This amount of gas could replace 1½–2 million barrels of imported oil a day. If this full utilization were realized, it would result in national balance-of-payment savings of $16.5–22 billion a year (at the current market price of $30 a barrel for imported oil).[2]

Pipelines are very efficient transporters of gas. By connecting gas wells with end consumers, pipelines avoid many of the problems plaguing competing fuels. The energy required for transmission is minimal, as gas flows readily under pressure.

Gas usage from residential and commercial customers for heating purposes tends to be seasonal. In order to bring about better utilization of the distribution network, the industry has established large storage facilities, which have a seasonal storage capacity of 7.3 trillion cubic feet of gas.[1] It is noteworthy that this storage capacity is equivalent to 1.2 billion barrels of oil, which is more than the strategic oil reserve planned by the federal government.

This storage capacity could be expanded to help the nation achieve a higher degree of energy security. A rational approach to strategic stockpiling would assign gas a major role in taking care of emergency needs for stationary heat and power. Such an objective could be accomplished at relatively modest cost. This procedure would reduce the excessive reliance on the federal government's Strategic Petroleum Reserve, which has been beset by many problems and cost overruns.[3] Mr. Eugene H. Luntey, president of the Brooklyn Union Gas Company, brought this matter to my attention. I believe it is in the national interest that government officials consult with Mr. Luntey and other gas industry leaders about this important subject.

The million-mile gas pipeline system and the 7.3 trillion cubic feet gas storage facilities are among the greatest assets this nation possesses in the struggle to achieve greater energy independence.[4]

Sources:
[1] Gas Facts 1978, *Department of Statistics, American Gas Association, 1979.*
[2] *"A.G.A.'s President Report," by George H. Lawrence, Financial Quarterly Review, American Gas Association, January 1979.*

3"*Washington's Ill-Starred Efforts to Stash Crude,*" *by Juan Cameron,* Fortune, *September 8, 1980.*

4"*Testimony before the Subcommittee on Energy and Power, U.S. House of Representatives,*" *by Henry Linden, president, Gas Research Institute, June 6, 1979.*

12. The Oil and Gas Connection

Throughout most of its history, natural gas was an adjunct to the oil business. Because oil was their primary interest, producers treated gas with considerable indifference. This attitude began to change with the emergence of national markets for gas as a result of the construction of interstate pipelines. Current technological and economic trends will reduce the linkage between oil and gas to a minimum.

Whenever oil is discovered, gas is very often present. Oil producers have frequently used gas as a driving force to help the flow of oil to the surface. In some instances, the gas pressure was so great that it interfered with the optimum recovery of the oil. Whenever that happened, the gas was vented into the atmosphere or burned off (flared). In effect, gas was treated as an adjunct to oil production. This situation still obtains in many parts of the world, where markets for gas have not as yet developed.

The first interstate gas pipelines were constructed in the 1930's. After the Second World War, a massive pipeline construction program took place, linking the gas-rich south-western regions of the U.S.A. with the energy-deficient north-east and midwest. This new economic role for gas enhanced its status to a more equal position. Ever since, gas has been considered a valuable fuel in its own right. Wells have been drilled for the deliberate purpose of finding gas. Currently, more than 70% of all the gas used in the U.S.A. is derived from gas wells, and less than 30% is produced in association with oil.

The emergence of the long-distance gas pipeline system had another profound effect on the relationship between oil and gas. In 1938 the U.S. government created the Federal Power Commission to regulate interstate commerce in gas. The oil companies feared that the Federal Power Commission would eventually expand its control to other areas associated with the pipelines. Therefore, the oil producers decided not to enter the gas pipeline field. The emergence of gas pipelines independent from oil companies greatly reduced the connection between oil and gas.

Current technological and economic trends will curtail even more the remaining linkage between oil and gas. From now on, drilling for gas will be increasingly separated from dilling for oil. As mentioned earlier, considerable effort is being devoted to the development of gas resources below 15,000 feet. At those depths, gas predominates because the temperatures are inimical to oil and favorable to gas. Similarly, Devonian shale, tight sands, and geopressured reservoirs contain primarily gas. Finally, gasification of coal and peat, as well as the manufacture of methane from waste products and other forms of biomass, involve construction of facilities independent of any connection with oil.

The independence of oil and gas serves the national interest. These major sources of energy can compete with each other in the discovery and development of new resources. Their efforts can facilitate the achievement of a high degree of energy self-sufficiency at optimum costs. This competition in the marketplace between oil and gas is a far sounder procedure for safeguarding the interests of consumers than the heavy hand of government regulators.

13. Why Gas is Cheaper than Oil

From the wellhead to the point of end use, natural gas enjoys significant cost advantages over oil. These economic realities provide gas with a secure future as a growing energy source.

Primary production from gas fields averages 70%, compared with 25–30% primary recovery from oil fields.[1] To increase production from oil wells, a variety of measures are employed, including injection of water and gas (secondary recovery), and the use of chemicals or heat (enhanced oil recovery). The employment of all these techniques may bring oil recovery up to 50%, still substantially less than primary gas recovery. Moreover, all efforts to increase oil output through external stimulation add to the cost of production.

Natural gas which is used in interstate commerce must meet pipeline quality specifications. If there are any impurities present, they are removed *before* the gas enters the pipeline. As a result, the need for expensive pollution control equipment at the point of end use is greatly reduced. Some types of fuel oil contain high amounts of sulfur and other pollutants, which may require treatment *after* the fuel is burned in order to comply with air quality standards.

One of the unique features of gas is its readiness for end-use applications without much further processing. In effect, natural gas has been refined to a considerable extent by nature. In contrast, crude oil is generally not used in its original state. It is usually processed in refineries and transformed into such products as gasoline, diesel fuel, or fuel oil before it can be used. These refining procedures add to the cost of the end products. It follows that realistic cost comparisons of gas should be made with refined products, not with crude oil.

The existing pipeline network for transporting gas from the wellhead to end users is the most cost-effective of all energy delivery systems. The only energy that is expended along the way is for compressors to push the gas along. In contrast, oil is generally distributed to end users by tank trucks, barges, or railroad equipment. Considerable amounts of energy are required for these procedures, which add to the cost of distribution.

The structure of methane, consisting of one carbon atom surrounded by four hydrogen atoms, gives it significant heating advantages over oil. According to a report by the Department of Energy/Federal Energy Regulatory Commission, the total-cycle efficiency for space heating ranges from 44–53% for natural gas, compared to 33–39% for fuel oil.[2] It takes approximately four units of oil to generate the same amount of heat as three units of gas.

The cost advantages of gas over oil cited thus far are applicable regardless of the oil's origins. In addition, gas can generate substantial savings to the national economy by reducing dependence on imported oil.

Gas is available in plentiful supplies from domestic and other nearby sources. As a result, the high risks and hidden costs of imported oil are avoided. About 44% of U.S. oil requirements

70

are dependent on imports, largely from countries outside the North American continent. These oil imports undermine the U.S. balance of payments, weaken the currency, contribute to inflation, cause unemployment, and add to national security costs.

A number of economists have attempted to quantify these hidden costs of imported oil. A thorough analysis of this matter was prepared by Professor Rod Lemon for the Institute of Gas Technology.[3, 4] Professor Lemon uses the term "external benefits" to designate the savings on hidden costs that would result from reduced oil imports. Here are his findings as of September 1, 1980:

Benefits of Reducing U.S. Oil Imports By 500,000 Barrels a Day[4]

	Year 1	Year 2	Year 3
	1980 $ per barrel		
Direct benefit*	37.00	37.74	38.49
External benefits			
Lower oil prices	12.32	17.71	18.98
Reduced inflation, improved trade, currency appreciation	11.19	20.41	21.18
Added real output, increased domestic employment	6.60	12.70	17.13
Decreased supply disruption costs	13.50	13.50	13.50
Subtotal of external benefits	43.61	64.32	70.79
Total benefits per barrel	80.61	102.06	109.28

*Based on higher world oil price scenario, and assuming annual price increases of 2% in real terms.

71

It is noteworthy that the external benefits from reduced oil imports are greater than the direct cost of the oil. To place this matter into perspective in relation to gas, let us see what 1,000 cubic feet of gas could cost to equal the above oil prices. One barrel of oil equals 6,000 cubic feet of gas. To arrive at the 1,000 cubic feet price, we divide the "total benefits" prices of the preceding table by 6, and get the following results: $13.43 for year 1; $17.01 for year 2; and $18.21 for year 3.

All types of gas, including shallow and deep conventional, Devonian shale, tight sands, geopressured, coal seams, gasification of coal, peat, and oil shale, as well as biomass conversion into methane, are substantially lower in price than the imported oil adjusted for external costs. On the basis of this scenario, national policy should encourage the maximum utilization of gas from all sources at the earliest possible time.

Sources:
[1]*John M. Hunt, The Future of Deep Conventional Gas," UNITAR Conference on Long-Term Energy Sources, Montreal, December 1979.*
[2]*"National Gas Survey: Efficiency in the Use of Gas," by the Department of Energy/Federal Regulatory Commission, June 1978.*
[3]*Rod Lemon, "The Direct and External Benefits of Reducing Oil Imports," Institute of Gas Technology, October 1, 1979.*
[4]Ibid., *"The Externalities of Oil Imports Revisited," Energy Topics, Institute of Gas Technology, September 1, 1980.*

14. A Rational Division of Markets Between Oil and Gas

The most economic use of oil is for the production of transportation fuels and chemical feedstocks. Natural gas is the ideal fuel for generating stationary heat and power. Improved utilization of these valuable resources for the tasks best performed by them would save the U.S. billions of dollars in oil import bills, would reduce costs to consumers, and would improve the environment.

When crude oil is refined into such transportation fuels as gasoline, diesel oil, and jet fuel, the most economic return on refinery investment is achieved. In contrast, the lowest return on investment is realized from residual fuel oil, which is marketed to industrial users and utilities for generating stationary power and heat. A recent study by Purvin & Gertz, Inc.[1] points out that the technology exists for increasing the refinery output of gasoline and diesel fuel and eliminating the residual fuel oil. An investment of $18 billion to upgrade U.S. refineries would yield additional gasoline supplies of 470,000 to 530,000 barrels a day as well as additional diesel fuel of 540,000–600,000 barrels a day. As a result, the U.S. import bill for foreign oil and refined products could be reduced by more than $12 billion annually. It is apparent that the complete conversion of crude oil into transportation fuels would serve the best interests of the oil industry and of the nation.

At least one oil company independently arrived at conclusions very similar to those presented in the Purvin & Gertz study. Ashland Oil announced plans to increase gasoline output per barrel of oil from 50% to 75% at its Ashland, Kentucky refinery. The process employed by Ashland is applicable to heavy, high-sulfur crude oil more readily available from domestic U.S. sources. Ashland will spend $70 million to construct the additional facilities.[2]

Natural gas is the cleanest and most efficient source of heat and power for stationary applications. Its environmental benefits are beyond dispute. In addition, the economic cost of natural gas compares favorably with competitive fuels, including residual fuel oil. The million-mile pipeline network that has been built to transport gas from producing fields to end users is highly efficient. However, its cost-effectiveness is to a large extent dependent on optimum loads. The better the balance among residential, commercial, and industrial customers, the more efficient pipeline operations become. A balanced pipeline load benefits all parties, including residential customers.

These economic realities are conducive to a more rational application of resources in the energy field. Oil companies should be urged to upgrade their refineries to produce more transportation fuels. Gas companies should be encouraged to supply the stationary markets previously using residual fuel oil. However, unsound government policies interfere with such rational behavior.

The gas shortages of the early 1970's, which were caused by the wrong government pricing policies for wellhead gas during

74

the preceding two decades, have caused the government to place restrictions on the sale of gas to industrial and utility customers. Many of these constraints are still in effect, even though currently there are no gas shortages. In fact, these counterproductive government policies jeopardize the expansion of gas production and gas utility operations. Moreover, the U.S. is spending billions of dollars for imported oil that could be replaced by domestic gas.

The national interest requires that the government act without delay in removing the obstacles which it has imposed on the marketing of gas to all customers the gas industry is ready to serve.

Source:
[1]Purvin & Gertz, Inc., "An Analysis of Potential for Upgrading Domestic Refining Capacity," May 1980.
[2]"Found: A Way to Squeeze a Barrel of Oil," *U.S. News & World Report*, April 14, 1980.

15. The Advantages of Heating with Gas

Natural gas compares favorably with oil for heating homes and offices. The clean-burning qualities of gas help maintain high equipment efficiency. Gas is continuously available from utility pipes connected directly to the end user. Breakthroughs in gas technology, notably the development of pulse combustion boilers and furnaces, can achieve unprecedented fuel efficiencies. Escalating oil prices in recent years have given gas significant cost advantages over oil.

Gas equipment maintains high operating efficiencies over long periods of time because of the fuel's clean-burning characteristics. In contrast, oil furnaces and boilers tend to lose efficiency with time. According to a study prepared for the National Bureau of Standards, oil furnaces may lose 1.9% per year in efficiency.[1]

The ready availability of gas from utility pipes connected to the residence or commercial building is a great advantage. Oil has to be periodically ordered and brought by truck to the customer.

The manufacturers of pulse combustion boilers and furnaces claim fuel efficiencies in excess of 90%. (See Chapter 10 for additional information). Thus far no oil-burning equipment of similar efficiency has been announced.

These features would suffice to assure gas significant market advantages, even if the price were the same as that of oil. However, the oil price escalations since 1973 have given gas a competitive edge on a cost basis as well. In effect, gas is a premium fuel selling at a discount from oil.

The following table illustrates the cost advantage of gas over oil for residential heating purposes.

Prices of Oil and Gas for Residential Heating[2]
(current $ per million Btu)

Year	Gas	#2 Fuel Oil
1960	1.00	1.08
1965	1.01	1.15
1970	1.06	1.33
1973	1.25	1.64
1975	1.69	2.81
1978	2.53	3.57
1980 (estimated)	3.45	5.22
1985 (estimated)	5.93	9.13
1990 (estimated)	9.48	13.16

By converting the preceding data into average annual heating bills, the savings for gas users become even more graphic.

Average Residential Heating Bills[2]
(current dollars)

Year	Gas	#2 Fuel Oil
1960	94	108
1965	106	129
1970	123	165
1973	140	196
1975	181	320
1978	242	372
1980 (estimated)	314	506
1985 (estimated)	492	806
1990 (estimated)	720	1,065

It should be noted that the preceding figures are averages for the nation as a whole. For more specific information, it is necessary to check with the utility serving a given market and setting prices within its regulatory framework. All estimates of future prices for both oil and gas involve considerable uncertainties.

The switch from oil to gas requires the purchase of new equipment, which ranges in cost from approximately $800 for a gas conversion burner, to $1,200 for a new gas furnace, and $1,900 or more for a new gas boiler.[3] Pulse combustion equipment prices are even higher. In order to get the best price, it is generally advisable to get more than one estimate for any given installation.

A study prepared by ICF, Inc. for the U.S. Environmental Protection Agency cites two circumstances particularly advantageous for converting from oil to gas:

(1) At the time of normal replacement, that is, when the oil furnace or boiler needs to be replaced in any case; and

(2) When the switch involves replacing oil equipment with a gas unit operating at higher efficiency. Fuel savings over a

number of years will make up the cost of the equipment.[4]

Apparently a great many consumers have reached the conclusion that it is in their best interest to convert from oil to gas. In 1979 alone over 350,000 such conversions took place. This trend is continuing in 1980.

The nation also gains from the replacement of oil furnaces and boilers with high efficiency gas equipment. These benefits include net savings of energy, lower oil import costs, and positive effects on the environment. The ICF study notes that "a small Federal program to encourage investment in high efficiency gas heating systems might lead to substantial amounts of energy savings with a low Federal budget cost per barrel of oil saved."[4]

It is to be hoped that some way can be found to facilitate the oil to gas conversion without imposing undue hardships on the oil jobbers who have hitherto served these markets. Most of these jobbers are independent businessmen who are caught in a predicament not of their own making. It would seem to be appropriate to compensate them in some fashion for their losses. The "excess profits" tax which the federal government collects from oil producers would be a suitable source of such compensatory payments.

Sources:
[1] *"A Study to Evaluate the Effect of Performing Various Energy Saving Procedures on Residential Oil Burner Installations in the New England Area,"* prepared by the Walden Research Division of Abcor, Inc. for the National Bureau of Standards, U.S. Department of Commerce, August 1975.
[2] *"Consumer Cost of Natural Gas and Alternative House Heating Fuels,"* Policy Evaluation & Analysis Group, American Gas Association, October 22, 1979.
[3] *"An Estimate of Costs and Payback Periods of Residential Oil-to-Gas Conversions,"* Ibid., *June 6, 1980.*

⁴*"An Analysis of the Economics of Replacing Existing Residential Furnaces and Boilers with High Efficiency Units," submitted to the Office of Planning and Evaluation, U.S. Environmental Protection Agency, by ICF, Inc., May 1980.*

16. The Future Looks Bright for Methane-Powered Vehicles

Compressed natural gas is an excellent fuel for transportation applications. It is considerably cheaper than gasoline or diesel fuel. It reduces engine wear and minimizes pollution. Disadvantages include the need for costly conversion equipment and the lack of a distribution network for servicing methane-powered vehicles. Under present circumstances, the latter are suitable primarily for fleet operators in urban areas.

The following comparisons, using 1979 figures, show the cost advantages of compressed natural gas:

1979 Alternate Fuels Costs for Vehicles[1]

Fuel	Cost (cents per mile)
Gasoline	4.7
Diesel	2.7
Compressed natural gas	1.8

Comparative data for 1980 are even more favorable for compressed natural gas. The cost of the latter is around 37 cents per

81

100 cubic feet, which approximates a gallon of gasoline in thermal content.[2] Meanwhile, gasoline prices have risen to more than $1.20 per gallon in most parts of the country.

In addition to the fuel cost savings, methane has the following advantages over gasoline:

(1) It reduces wear and maintenance on the engine because of its clean-burning characteristics;

(2) Spark plugs last longer;

(3) Oil changes are less frequent because of reduced contamination;

(4) Because methane is a gas, it enables the car to start more easily in cold and hot climates; and

(5) It reduces air pollution.[3]

The environmental advantages of methane over gasoline are striking. The following table compares air pollutants generated by these fuels.

Air Pollution Emissions (grams per mile)[4]

Pollutant	Methane	Gasoline
Hydrocarbon	0.26	0.43
Carbon monoxide	0.09	2.72
Nitrous oxide	0.50	1.65
Sulfur oxide	negligible	0.71
Particulates	negligible	0.08

Methane exhaust products are non-reactive in photochemical reactions. As a result, they do not form smog. The high octane rating of methane (130) avoids the use of additives, such as lead, which tend to be harmful to automobile engines and to people.[1]

Methane is also safer than gasoline or diesel fuel. In case a methane container leaks, the methane, being lighter than air, tends to dissipate harmlessly into the atmosphere, whereas gasoline or diesel fuel, which puddles on the ground, may pose a fire hazard. Other safety features of methane include its high ignition temperature (300–400° Fahrenheit higher than that of liquid fuels) and the narrow limits within which air and methane must mix if combustion is to take place.[1]

Practical experience supports the value of these safety features. A study prepared by Dr. J. Winston Porter shows that in over 175 million miles driven by methane-fueled vehicles in the United States since 1970, no deaths and only one injury occurred in which natural gas was a contributory factor. In an estimated 1,360 collisions involving such vehicles, including 180 rear-end collisions, there were no failures or fires involving the methane system.[5]

The main impediments to conversion from gasoline or diesel to methane include:

(1) The cost of conversion;

(2) Lack of an infrastructure to provide methane fuel; and

(3) Range restrictions.

Until automobile manufacturers produce cars specifically designed for methane, conversion equipment must be added to existing vehicles. The cost of this conversion is $1,300–1,800 per vehicle.[4] According to the manufacturer of the conversion equipment, sales have risen sharply since the 1979 gasoline shortages and price increases.[3] The higher gasoline prices go, the more attractive conversion to methane becomes.

At the present time, there is no national network for providing methane service. Under proper economic conditions, such a network could be established. The million miles of natural gas pipelines reaching most parts of the country would facilitate setting up distribution outlets.

Range limitations are surmountable by adding an extra compressed fuel tank to the vehicle. If the automobile or truck were designed for methane in the first place, sufficient gas-holding

83

capacity could be included to avoid range limitations.

Under present circumstances, methane-powered vehicles are suitable primarily for urban fleet operations, which have their own methane service facilities. Taxi fleets, utility vehicles, and similar mass users of automobiles and light trucks would find it worthwhile to explore this alternative to high-cost gasoline power.

There are currently some 6½ million vehicles in fleet operations. They consume 700,000 barrels of oil products a day. The gas industry has the capacity to make inroads in this important market.[6]

It is to be hoped that the U.S. Department of Energy and other government agencies will devote more attention to the use of methane as an alternative transportation fuel. This approach makes more economic sense than the other options currently available, including electric cars.[1, 4] Moreover, government agencies and the post office should make increased use of this cost-saving and environmentally beneficial approach.

Sources:
[1]*Benjamin Schlesinger, Nelson E. Hay, and Michael I. German, "Methane Powered Vehicles: Short and Long Term Potential," American Gas Association, February 1980.*
[2]*"CNG Vehicle Test Successful; Chicago Utility Adds 50 More,"* Gas Industries, *September 1980.*
[3]*"CNG Carburetion Sales Gain,"* ibid.
[4]"A Comparison of Alternative Synthetic Fuels for Light Vehicles: 1980 Update," Policy Evaluation & Analysis Group, American Gas Association, June 11, 1980.
[5]*Dr. J. Winston Porter, "Preliminary Analysis of the Safety History of Natural Gas-Fueled Transportation Vehicles," Gas Energy Review, American Gas Association, February 1980.*
[6]*"A.G.A. Promotes Synfuels with 'Coal Gas Van' ", American Gas Association.*

17. Coal and Gas—Compatible Partners

The national interest requires that the coal and gas industries work together to quicken the pace of U.S. energy independence. The conversion of electric utilities from oil to coal can be facilitated by the select use of gas to help meet environmental standards. Gas can also fulfill the function of providing energy for meeting peak load requirements by utilities using coal as the basic fuel. The gas industry may soon become a major market for methane from coal seams and for coal for gasification projects.

The Powerplant Fuels Conservation Act of 1980 passed by the U.S. Senate provides that 51 electric powerplants currently using oil should be switched to coal as soon as possible. Environmental considerations pose a serious challenge to this program. When burned, coal emits large quantities of sulfur dioxide and particulates, which would be in violation of air quality codes in many places. The problem can be solved by the select use of natural gas, which is virtually free from sulfur and particulate emissions. According to an analysis prepared by the

85

American Gas Association, a combination of 23–38% natural gas with 77–62% coal would prevent any increase in sulfur dioxide emissions from current oil-generated levels.[1] Overall, the use of 280–460 billion cubic feet of natural gas (a 1–2% increase in U.S. gas use), would make possible the combustion of an additional 34–42 million tons of coal a year (a 7% increase in coal use) without adding to sulfur dioxide emissions. This joint effort by coal and gas would replace 576,000 barrels of oil a day, providing an annual balance of payment saving of $6.3 billion (when oil is calculated at $30 a barrel).[1]

In addition to solving the environmental problem, gas can also help to improve the operating performance of electric utilities by providing fuel for peak load requirements. For example, during hot summer days the demands on electric power facilities are considerably above normal because of heavy air conditioning use. Such peak loads can be met without costly capital expenditures by using gas. This clean fuel will be particularly welcomed on sweltry days, when the population is under stress from the heat and any reduction in air pollution brings great benefits to well-being.

The recovery of methane from coal seams can help make coal mining safer and more profitable, while providing a valuable addition to the nation's energy supplies. Similarly, low-grade types of coal and coal situated in seams that are uneconomic to mine may find markets via the gasification route. It is likely that gas utilities will become increasingly important customers of the coal industry. A single coal gasification facility, like the Great Plains Coal Gasification Project currently under construction in North Dakota, will use over 8 million tons of lignite coal a year.[2] It may well be that before the end of this century gas utilities will be the biggest customers of the coal industry.

This analysis indicates that the coal and gas industries have much to gain from close collaboration. Every effort should be made to facilitate the development of this compatible partnership, which is also in the best interest of the nation.

Sources:
[1] *"Analysis of Select Gas Use in Utility Coal Conversion for Limiting Sulfur Emissions," Policy Evaluation & Analysis Group, American Gas Association, April 29, 1980.*
[2] *"The Great Plains Story,"* Venture, *Summer 1980 (published by American Natural Resources System).*

18. Gas Keeps the Environment Clean

Natural gas is the cleanest of all fossil fuels. Its production involves minimal disturbance of the surroundings. Cleaning of the gas *before* it is put into pipelines removes most pollutants. When gas is burned, it combines with oxygen to form water and carbon dioxide, two harmless substances. Gas is worthy of support by all those concerned with the environment.

Drilling for natural gas is basically a clean operation, involving very little permanent disturbance of the surrounding area. Similar comments apply to gas production procedures. There are no unsightly structures, nor is there any large-scale removal of plants or soil. Once production has been completed, the drilling installations are removed and the environment is quickly restored to its original state.

To meet the standards of pipeline quality, virtually all pollutants are removed from the natural gas *before* it is allowed entry. This procedure differentiates gas from other fossil fuels. By the time gas reaches the consumer, it is as clean as any fuel can be. In contrast, with most other fossil fuels the attempt is made to remove pollutants *after* they are burned, a procedure that is inherently less effective.

The actual burning of natural gas, which is primarily methane, is very simple and clean. One methane molecule combines with two oxygen molecules to form two water molecules and one carbon dioxide molecule. The chemical shorthand for this reaction is $CH_4 + 2O_2 \rightarrow 2H_2O + CO_2$. The water molecules ($H_2O$) are in the form of vapor, which will eventually return to the ground in the form of rain or dew. The carbon dioxide generally serves as food for plants.

The cleanliness of gas is illustrated by the following table, which compares gas with oil and coal. The data for this table are based on information provided by the Environmental Protection Agency,[1] the U.S. Department of Energy,[2] and Hittman Associates, Inc.[3]

Pounds of Air Pollutants per Billion Btu

	Gas	Oil	Coal
Sulfur oxides	0.6	830–920	660–4,390
Particulates	5–15	140–720	60–9,440
Carbon monoxide	17–20	40	44–88
Hydrocarbons	1–8	7	13–44
Nitrogen oxides	80–700	130–760	670–2,440

A comment about nitrogen oxides may be in order. Nitrogen is a constituent of air and will form compounds with oxygen (nitrogen oxides) whenever heat is generated by any type of combustion in the presence of air. As the above table shows, even in this category gas has some environmental advantages over the other fuels.

Gas deserves the active support of all individuals who are concerned with improving the environment. It is essential that this support be extended to exploration and production of gas, as well as to its end use. Adequate supplies are necessary if gas is to achieve its beneficial effects for the environment.

Sources:
[1]*"Complilation of Air Pollutant Emission Factors," Third Edition, Environmental Protection Agency, May 1978.*
[2]Monthly Energy Review, *May 1978, U.S. Department of Energy.*
[3]Environmental Impacts, Efficiency, and Cost of Energy Supply and End-Use, *Volume 1, Hittman Associates, Inc. for the National Science Foundation, the Environmental Protection Agency, and the Council on Environmental Quality, Columbia, Maryland, November 1974.*

19. The High Cost of Government Price Manipulations

The Federal Power Commission's imposition of artificially low prices on gas producers since the 1950's has seriously hampered the development of adequate gas supplies in subsequent years. Much of the recent energy crisis may be attributed to the federal government's irrational economic behavior.

In 1954 the Supreme Court ruled in the Phillips Petroleum case that the Federal Power Commission had to regulate the wellhead price of gas paid by interstate pipeline systems to independent producers. On the basis of this ruling, the Federal Power Commission set prices based on historical costs for finding and processing gas. These prices were too low to allow adequate exploration for new sources. (See end of chapter for specific prices).

Whenever the price of an economic good or service is kept below its true value by government fiat, several consequences ensue. The demand for the product is stimulated beyond normal levels. At the same time, the availability of the product tends to decline, as producers have little incentive to keep up the search for new sources of supply. It is inevitable that sooner or later shortages will develop.

All of these consequences were demonstrated as a result of the actions by the Federal Power Commission to control well-head gas prices. The demand for gas increased steadily between 1954 and 1973.[1] However, the incentive for producers to find and develop new sources of gas supply was reduced. After the mid-1950's, drilling activity dropped significantly and by 1968 estimated proved reserves started to decline. Between 1968 and 1979, proved gas reserves were reduced from 293 trillion cubic feet to 195 trillion cubic feet.[2]

This decline in estimated proved reserves necessitated curtailment in the sale of gas to electric utilities and industrial customers by the early 1970's. Most of the gas was replaced by imported oil. This development increased U.S. dependence on imported oil at a time when the price of the latter was escalating.

The proponents of government gas pricing policies claim that consumers benefit from the low cost of gas. In actuality, the explosive price increases in energy during the past few years have demonstrated the fallacy of this claim. Irrational economic behavior by the government does not serve the best interests of consumers or producers. The overall record indicates that the American people would have been better off if the government had stayed out of the price manipulation business.

It is important to keep in mind that the gas shortages experienced by the U.S.A. in the 1970's were largely, if not solely, due to wrong government policies. Domestic gas resources are plentiful, but there must be economic incentives to justify their utilization. We must avoid a situation in which the government compounds its erroneous behavior by placing additional obstacles in the way of the free market. Once the shackles of counterproductive government policies are removed, the U.S.A. can once again achieve a high degree of energy independence and a prosperous economy.

Average Gas Prices at the Wellhead[1]
($ per thousand cubic feet)

Year	Price	Year	Price
1955	0.104	1971	0.182
1956	0.108	1972	0.186
1957	0.113	1973	0.216
1958	0.119	1974	0.304
1959	0.129	1975	0.445
1960	0.140	1976	0.580
		1977	0.790
1961	0.151	1978	0.899
1962	0.155		
1963	0.158		
1964	0.154		
1965	0.156		
1966	0.157		
1967	0.160		
1968	0.164		
1969	0.167		
1970	0.171		

Sources:
[1]*Gas Facts 1978,* American Gas Association.
[2]*Reserves of Crude Oil, Natural Gas Liquids and Natural Gas in the United States and Canada as of December 31, 1979,* American Petroleum Institute, June 1980.

20. The Federal Government Finally Moves in the Right Direction

It took the federal government five years after the onset of the acute energy crisis in 1973 to pass meaningful legislation that would undo some of the damage resulting from previous government actions. By moving in the direction of more realistic pricing procedures, the Natural Gas Policy Act of 1978 has already resulted in sharply stepped-up exploration for and production of gas. This experience demonstrates that increased reliance on the free market will bring about the best solution to the energy problem.

The Natural Gas Policy Act of 1978 provides for the gradual decontrol of new gas prices by 1985. In addition, as of December 1979 it exempted the following types of gas from price controls:

(1) Gas formations below 15,000 feet;
(2) Geopressured gas;
(3) Gas from coal seams;
(4) Gas from Devonian shale.

It also established a special incentive price system for gas which can only be produced under conditions involving extraordinary risks or costs, as determined by the Federal Energy Regulatory Commission.[1]

While this legislation has by no means removed all the impediments to the maximum production of gas from domestic sources, it marks a significant departure from the sterile and counterproductive orientation that had previously characterized government policies. The response by gas producers provides strong evidence that the federal government is finally moving in the right direction. Drilling activity for new gas has increased dramatically, reaching near-record levels in 1979 and continuing at a high pace in 1980. Additions to proven gas reserves increased by 35% in 1979 over 1978.[2]

As a result of these positive supply developments and conservation, the gas industry is well positioned to serve existing markets and to expand its activities to additional consumers. In 1979 the gas industry added more than 350,000 new customers who switched from oil to gas for space heating in homes. In is expected that additional hundreds of thousands of residential customers will join the ranks of gas users in 1980 and subsequent years.[3]

The gas industry has also been able to improve its service to industrial customers, though its marketing activities in that area are still seriously hampered by government restrictions. It is estimated that gas utilities sold enough additional gas to industrial customers in 1979 to offset foreign oil imports at the rate of 435,000 barrels a day. This contribution by the gas industry saved the U.S.A. approximately $4 billion in 1979 foreign oil payments.[4] It is noteworthy that the gas industry officials expressed the opinion that they could have offset 1.2 million barrels of daily oil imports if there had been no constraints of any kind on sales to industrial customers.[4] It is evident that gas utilities and all of their customers would benefit from increased gas production and less government interference in marketing activities. These comments apply to both federal and state regulatory authorities.

The impressive results achieved by the gas industry since the passage of the Natural Gas Policy Act of 1978 provide strong evidence that the free market approach works best in dealing with the energy problem. It may be worthwhile to look at this reality more closely.

The free market is not just some abstract concept dreamed up by a few economists. In the case of gas, the free market involves the motivations and activities of hundreds of thousands of people who possess the greatest skills in the production, transportation, marketing, and distribution of gas to millions of end users. Life-long careers have been dedicated by these professionals to the task of doing the best possible job. Free markets, including unregulated prices, are essential if optimum performance is to be accomplished. The historical record clearly shows that the free market system works well, and benefits all parties, particularly consumers.

In contrast, whenever the government has taken over control of prices, marketing, or other facets of the gas business, the results have been unfortunate, not only for the gas industry, but also for the consumers and the country. We may grant that the government has good intentions. However, as the saying goes, "the road to hell is paved with good intentions." Even if government officials were very intelligent and had the best interests of the country at heart, they could not do as good a job as the free market. The latter is an organic reality, involving the economic decisions of millions of people, whose interaction results in an optimum allocation of resources. On the other hand, the fiat of government officials creates an artificial situation that leads to fundamental distortions which can undermine a whole industry. Recent history has demonstrated the validity of this assessment in the energy field. We can only hope that everyone concerned with the welfare of the country will learn the appropriate lessons from this experience.

Sources:
[1] *The Natural Gas Policy Act of 1978, signed into law on November 19, 1978.*
[2] *"Proved U.S. Reserves—Year-End 1979," Gas Supply Committee, American Gas Association, Gas Energy Review, June 1980.*
[3] *"An Estimate of Costs and Payback Periods of Residential Oil-to-Gas Conversions," Policy Evaluation & Analysis Group, American Gas Association, June 6, 1980.*
[4] *"Survey of Actual 1979 Industrial Oil Offsets and Potential Offsets in the First Half of 1980," Policy Evaluation & Analysis Group, American Gas Association, December 21, 1979.*

21. Government Gas Marketing Restrictions Should be Lifted

Due to government-mandated marketing restrictions imposed in the 1970's, the gas industry has lost sales aggregating some three trillion cubic feet a year to industrial and electric utility customers. About two-thirds of this gas has been replaced with imported oil.[1] At current prices of $30 per barrel of oil, the cost to the nation is over $16 billion in foreign exchange. Now that gas supplies are once more available in adequate amounts, gas companies should be freed to go after this market without further delay.

In the early 1970's the U.S. experienced gas shortages. As has been shown previously, these shortages were caused by wrong government policies in relation to wellhead prices of gas. To deal with these shortages, the government imposed restrictions on the sale of gas to industrial and electric utility customers. This market loss has had adverse effects on the gas industry as well as on the national economy.

The gas industry requires a balanced mix of customers to function economically. Residential and commercial consumers

98

use gas primarily for heating. Their demand for gas is at a peak in the winter months, and falls off sharply during the rest of the year. If the gas industry were confined to these markets, its capital investment would be uneconomically utilized, resulting in a high-cost operation. Industrial and electric utility customers improve the overall utilization of the gas industry's facilities, thereby lowering costs for all customers.

To illustrate this point, many electric utilities experience their highest demand during the hot summer months, because of the widespread use of air conditioners. Gas consumption is at a low ebb during the summer. It is therefore to the mutual benefit of both the gas industry and the electric utilities to make use of excess gas supplies during the summer for powerplant operations. This approach has the additional advantage of keeping air pollution levels down during the uncomfortable summer months.

Gas can also facilitate the increased use of coal to replace imported oil in powerplants, by keeping overall pollution levels down. Similarly, gas is an ideal fuel for helping utilities meet peak load requirements.

Industrial customers generally provide a sustained, year-round market to gas companies. In addition, some industries can switch to gas when the latter is in plentiful supply and go back to other fuels when gas is needed for other priority customers. In either case, industrial customers will contribute to improved utilization of gas resources.

Balanced and sustained markets, with a variety of customers, are essential to the gas industry if it is to function at optimum levels. Moreover, the expansion of facilities, and the ambitious plans for adding to gas supplies, are meaningless without the ability to compete for ever-growing markets for gas. It is unlikely that the gas industry will spend hundreds of billions of dollars on gas production, transportation, and utility facilities if the government undermines economical operations with outdated marketing restrictions.

The national interest requires maximum use of gas for all applications for which it is best suited. With the increased supplies of gas now becoming available, companies in the gas industry should be allowed the right to determine their own marketing strategies, without the heavy hand of government interference. It is a reasonable assumption that each gas company knows its own resources and markets better than the bureaucrats and politicians in Washington and in the state capitols. Marketing decisions by individual companies are far more likely to be based on sound economic principles than those imposed by government fiat.

Sources:
[1]*"Natural Gas Deliveries, Curtailment, and Alternate Fuel Offsets of Curtailment, April 1974 through March 1977," Federal Energy Regulatory Commission, May 1978.*

Other sources:
    "Critique of Department of Energy's Interim Rules to Implement the Powerplant and Environmental Fuel Use Act of 1978," Report of the Regulatory Analysis Review Group, U.S. Government, Oct. 31, 1979.
    Wharton Econometric Forecasting Associates, Inc., "National Economic Impacts of Natural Gas Policy Act Incremental Pricing: 1980 Update," February 1980.
    Energy Analyses prepared by the Policy Evaluation & Analysis Group, American Gas Association, dated March 30, 1979, April 11, 1979, May 9, 1979, and July 6, 1979. All of these studies deal with the impact of incremental pricing on the gas industry.

22. Gas Can Help Prevent Disaster

Excessive dependence on imported oil poses serious hazards to the U.S.A. Recent analyses by the Wharton Econometric Forecasting Associates and by the American Gas Association indicate that a complete disruption of oil imports from the Persian Gulf region could add 8,600,000 people to U.S. unemployment rolls and could reduce the gross national product by $350 billion.[1] These disastrous consequences could be largely prevented by the increased used of gas and coal, together with improved conservation.

The free world depends on the Persian Gulf region for 60% of its total oil imports. Most of this oil has to move through the Straits of Hormuz, a passageway highly vulnerable to disruption. If this supply route were closed for a lengthy period, it could have a devastating effect on the world economy, including that of the U.S.A. Unemployment would soar, gross national product and personal income would decline sharply, and other economic indicators would also go into a tailspin.[1]

Most of these adverse effects could be ameliorated or prevented altogether by the increased use of gas, coal, and conservation.[2] About 80% of all U.S. oil imports, or 5.5 million barrels a day, are used in large stationary applications, such as electric powerplants and industry. A combination of gas and coal could replace that oil.

The following chart was prepared by the American Gas Association to show how 5.5 million barrels of oil per day could be displaced by gas and coal. "Category One" refers to oil for which gas could be substituted almost immediately. "Category Two" oil displacement requires a somewhat longer time span and/or the use of gas and coal together.

Oil Use Replaceable By Gas and Coal[3]
(million barrels per day crude oil equivalent)

	Category 1	Category 2	Total
Industry			
Liquefied Petroleum Gas (LPG)	—	.4	.4
Residual oil	.4	.3	.7
Distillate	.1	.5	.6
Subtotal	.5	1.2	1.7
Power plant			
Residual oil	.6	1.0	1.6
Distillate	—	.1	.1
Subtotal	.6	1.1	1.7
Residential and commercial buildings			
Residual oil	—	.5	.5
Distillate	—	1.6	1.6
Subtotal	—	2.1	2.1
TOTAL	1.1	4.4	5.5

Altogether, this program would involve the use of an additional 190 million tons of coal and 3.4 trillion cubic feet of gas per annum.

The high rate of exploration and drilling for gas currently taking place in the U.S.A. can make a major contribution to the realization of the above program. In addition, the gas industry is developing the following supplemental gas supplies. The data are presented in terms of thousands of barrels of oil per day.

Near-Term Supplemental Gas Supply Potential[4]
(thousand barrels per day of oil equivalent)

	Estimated Potential by 1985
Canadian gas	500
Mexican gas	500
Alaskan pipeline gas	500
Liquefied Natural Gas (LNG)	500
Coal gasification	100
Synthetic natural gas	100
Landfill gas	100
Total	2300

Within this framework, the gas industry has the capability to help ward off a potential oil supply disaster. Moreover, existing gas pipelines can carry most of this gas to the consumers who need it. In addition, new pipelines are expected to be constructed soon to link up gas fields in Alaska and in the Rocky Mountain Overthrust Belt with consuming markets.

A study prepared by the Mellon Institute entitled "The Least-Cost Energy Strategy,"[5] reveals that conservation and the use of lowest cost energy sources could save consumers large amounts of money, while reducing overall energy consumption.

For example, if the proper energy policies had been followed over the past ten years, oil consumption would have been 30% less than it was in 1978. Realistic conservation measures should play a key role in all sound energy planning.

A strong case can be made that the financial drain resulting from excessive oil imports constitutes a disaster in and of itself. The estimated $90 billion the U.S. is paying for imported oil in 1980 has contributed significantly to inflation and has helped precipitate a recession leading to the loss of millions of jobs. Dependence on imported oil results in a "no-win" situation: the nation suffers disastrous consequences from paying the bills for supplies that reach its shores as well as from potential disruptions of imports.

Conservation, gas and coal can help extricate the nation from the terrible dilemma in which it currently finds itself. All possible steps should be taken to increase production of gas and coal to replace imported oil in providing fuel to industry and to electric powerplants. Simultaneously, a maximum effort should be made to conserve fuel through improved maintenance of equipment, the elimination of waste, the utilization of energy-saving devices, and the conversion to more frugal lifestyles. While this approach will entail some readjustments and sacrifices, it is infinitely preferable to a continuation of the present predicament.

Sources:
[1] *"Potential of Increased Gas Supply Capability to Reduce Impacts to the U.S. Economy of a Major Oil Supply Distruption,"* Policy Evaluation & Analysis Group, American Gas Association, June 27, 1980.
[2] *"Written Comments of the American Gas Association to the U.S. Department of Energy to Assist in Developing a Plan for Substantially Reducing by 1985 U.S. Vulnerability to an Oil Import Disruption,"* George H. Lawrence, Michael I. German, and Donald J. Schellhardt, May 30, 1980.

[3]*"Potential Substitution of Oil with Gas and Coal in Non-Transportation Uses," Policy Evaluation & Analysis Group, American Gas Association, August 28, 1980.*

[4]*"Briefing on Gas Energy and Contingency Planning," Policy Evaluation & Analysis Group, American Gas Association, Sept. 3, 1980.*

[5]*"The Least-Cost Energy Strategy," by Roger W. Sant, The Energy Productivity Center, Mellon Institute, Arlington, Virginia, 1979.*

23. Over $300 Billion for Gas Industry Expansion

The gas industry is poised for vast expansion programs. Capital requirements for the remaining two decades of this century are estimated at over $300 billion.[1] It is noteworthy that about 70% of these expenditures ($215 billion) will be devoted to the development of new gas supplies. This focus on assuring adequate gas resources to meet anticipated demand is a new departure for an industry that has historically depended on outside suppliers for its product.

The table below provides information on how the gas industry expects to allocate these capital funds.[1]

Estimate of Capital Requirements
For the U.S. Gas Utility Industry
1979–2000

	Billions of 1978 dollars
Conventional gas exploration and development in the continental U.S.A.	85.3
Supplemental sources	
Coal gasification	58.7
Alaskan gas	30.2
New technologies (unconventional gas)	30.2
Liquefied natural gas (LNG) imports	10.2
Mexican gas	0.2
Synthetic natural gas from light hydrocarbons	0.1
Total new gas supplies	214.9
Utility and pipeline construction and maintenance	87.6
Total capital requirements	302.5

The preceding figures include only the estimated expenditures by U.S. gas utilities. Some of these projects will also involve substantial investments by others.

In the 1970's the gas utilities experienced serious shortages of gas supplies. As a result, they had to curtail services to some of their customers, particularly in the industrial and electric utility field, and their marketing activities came to a virtual standstill. This experience provides the background for the decision by many gas pipeline and utility companies to assure themselves of adequate supplies in the future through their own efforts. It should be noted that even with this ambitious program, the gas industry would continue to purchase the bulk of its conventional gas requirements from outside suppliers.

By participating in exploration activities, the gas industry will not only develop assured sources of supply, but will also gain experience in such high-technology, capital-intensive endeavors as drilling for deep gas (below 15,000 feet), off-shore exploration, and drilling in difficult environments, including the arctic areas of Alaska and the Rocky Mountain Overthrust Belt.

The contributions by supplemental sources to U.S. gas supplies are expected to be substantial, as is evident from the following table. It should be emphasized that this forecast is subject to change. Technological and/or economic developments may warrant a shift of investments from less promising to more attractive areas. Fortunately the gas industry has many options for investing its funds, including conventional and unconventional sources of gas, domestic and foreign. No one should overlook the importance of capital in determining the future flow of gas supplies.

Potential Contributions from Supplemental Sources of Gas[1]
(trillion cubic feet)

Source	Forecast		
	1980	1990	2000
New technologies	0.05	1.8	5.0
Alaskan gas	—	1.6	3.6
Coal gasification	—	0.6	3.3
LNG imports	0.4	2.0	3.0
Canadian imports	1.4	1.1	0.8
Mexican imports	0.2	1.0	1.0
SNG from light hydrocarbons	0.5	0.5	0.5
Total	2.5	8.6	17.2

For the year 2000, the estimates for new technologies (unconventional sources) include 2.0 trillion cubic feet from tight sand formations, 1.2 trillion cubic feet from geopressured aquifers, 0.9 trillion cubic feet from Devonian shale, 0.3 trillion cubic feet each from peat gasification and from solid waste gas recovery, 0.2 trillion cubic feet from biomass conversion, and 0.1 trillion cubic feet from methane locked into coal seams. Coal gasification is treated separately. The $30 billion to be spent by the gas utility industry for new technology gas would be committed to development and production; it is assumed that the basic research will be funded primarily by public sources.

If major technological breakthroughs occur, or if economic conditions favor significantly higher gas prices, unconventional sources of gas could make even greater contributions to U.S. gas supplies by the year 2000. If such increased activities in new technology occur, capital requirements may have to be raised.

The North Slope of Alaska is expected to be serviced by one pipeline in the mid-1980's, to be followed by a second pipeline or added capacity in the initial pipeline in the 1990's. Gas from southern Alaska will be liquefied and transported in LNG tankers to the west coast of the continental U.S.A. The exact timetable of pipeline construction is still somewhat uncertain. The enormous proven and potential gas resources in Alaska can play an important role in meeting U.S. requirements for many years.

One commercial coal gasification facility, the Great Plains Coal Gasification Plant in Mercer County, North Dakota, is currently under construction. It is expected that by 1990 the U.S. will have eight such facilities, and by 2000 about forty may be operational. The $58.7 billion capital requirements estimated for coal gasification projects do not include $6.3 billion estimated capital costs for the coal mines supplying the raw materials.

If all the LNG projects in the planning stage are approved and completed, total capital expenditures will aggregate $36.5 billion, of which the U.S. gas utility portion is $10.2 billion. The balance of $26.3 billion will be spent by the foreign suppliers for liquefaction and terminalling facilities ($20.7 billion) and for tankers ($5.6 billion).

Imports of gas from Canada and Mexico require very little additional capital investment by U.S. gas utility companies. The gas from these sources is generally connected to existing U.S. pipelines, or to new ones being constructed (e.g., the Alaskan pipeline). In view of the enormous gas resources of both Canada and Mexico, the projected import figures appear unusually modest.

The $87.6 billion to be invested in utility and pipeline construction and maintenance during the next two decades is in line with current investments in this area. It should be noted that the amounts involved are actually larger than the total book value of such facilities (approximately $65 billion). The book value grossly understates replacement value. The gas utilities and pipelines are an invaluable asset to the nation.

The preceding data on capital requirements and applications are based on the gas utility industry's own estimates. The industry apparently assumes that demand for gas by the year 2000 will be approximately 30 trillion cubic feet, of which about 17 trillion cubic feet will be supplied by supplemental sources and 13 trillion cubic feet by conventional gas. It is noteworthy that this scenario projects a one-third decline in conventional gas supplies from the 19 trillion cubic feet level currently prevailing.

It may be presumptuous for an outsider to challenge the industry's projections, but for the sake of stimulating additional thinking on this matter, the author offers the following comments. The industry's estimate of total gas requirements for the

110

year 2000 appears to be too modest. Assuming that the industry is allowed a free hand to market its products, it is possible that the demand for and supplies of gas may grow to as much as 40 trillion cubic feet by the year 2000. Such an optimistic appraisal is in line with a generally growing conviction that gas is an ideal fuel and will emerge as the new energy leader.

Moreover, the author has the impression that the gas utility industry may be underestimating the conventional gas supplies that may become available over the next two decades. The pessimistic outlook for conventional gas by the industry may be partly due to the shock effects of the gas shortages experienced during the 1970's. As indicated elsewhere, these shortages were caused by unsound government pricing policies of gas at the wellhead. Assuming that these mistakes will not be repeated, it is likely that conventional gas supplies will be able to hold their own and may even increase over the next twenty years. Recent discoveries of enormous quantities of gas below 15,000 feet support this viewpoint. It would thus appear that most, if not all, of the extra 10 trillion cubic feet which is projected as a possibility for the year 2000 can be met from increased conventional gas supplies, supplemented by additional imports from Canada and Mexico.

If this assessment is correct, the gas utility industry should probably make a greater effort in conventional gas exploration and development than is currently contemplated.

Can the gas utility industry raise the large sums of capital required to meet its objectives? To help place the problem into clearer perspective, it is noteworthy that under this program the share of capital required by the gas industry will rise from its current 2% of U.S. gross annual private investment to almost 5% in the 1990's.[2] It is self-evident that the industry must be financially strong if it is to attract these enormous amounts of capital. All levels of government, federal, state, and local, must improve the regulatory and economic environment to facilitate this objective.

Regulatory agencies have a tendency to make their rate decisions on the basis of plant investments already in place. This approach is totally inadequate under the circumstances now prevailing. Inflation has undermined the value of traditional depreciation reserves. The large new expenditures, and the higher risks that are entailed, must be properly recognized if the industry is to realize its objectives. Similarly, the rate-making process must be speeded up, lest the industry lose opportunities for making timely decisions.

It should be remembered by state and federal regulators that the production of gas is a different business from its transmission and distribution. Costs of exploration, start-up expenses for new production facilities, personnel training programs, and other expenses connected with these new ventures should be deductible from current income. The criteria employed should be similar to those applicable to independent oil and gas producers. One possible way of accomplishing this objective is to apply separate treatment to gas production and the distribution business. The highly competitive nature of the production field makes regulation unnecessary and even counterproductive.

Higher gas prices to consumers are likely in the future, whether the gas utility industry integrates production capabilities into its business or not. It is far better for all parties, including the regulatory agencies, to face these realities now than to postpone the day of reckoning to a time when lack of advance planning leads again to serious gas shortages and/or exorbitant price increases. It is to be hoped that government officials have learned this lesson from the last quarter century's experience !

If realism and pragmatic compromise prevails among gas utilities, regulatory agencies, the federal government, and

112

consumers, there is little doubt that the investment community will provide the funds needed by the gas industry to accomplish its goals. Moreover, foreign investors are likely to be attracted to opportunities in the U.S. gas market. Europeans and Japanese financial specialists appear to have a greater awareness of the outstanding prospects for gas than their American counterparts.[3]

Sources:
[1] *"A Forecast of Capital Requirements of the U.S. Gas Utility Industry to the Year 2000," Policy Evaluation and Analysis Group, American Gas Association, April 20, 1979.*
[2] *"Future Capital Needs Now Pegged at $303 Billion," Financial Quarterly Review, American Gas Association, July 1979.*
[3] *"Overseas Financing," presented to an AGA conference on May 2, 1980, by John W. Barr of Morgan Stanley & Co.*

24. The Gas Industry Must Plan for Growth

During the next two decades, the gas industry will experience the biggest challenges of its history. To achieve the ambitious goals for growth, managements will have to mobilize all their talents. With the proper input of human skills and financial resources, most obstacles to growth can be overcome. The unique features of gas, its potential for energy leadership, and the strong marketing position of the gas utilities should facilitate the successful accomplishment of the tasks ahead.

Over the next twenty years, the gas industry will not merely increase in size, but it will also grow in complexity. Many utilities will be transformed from mere distributors of gas to integrated energy resource companies. The latter will eventually own or control a substantial part of their gas supplies. In addition to exploring for and producing gas, they may participate in coal and peat gasification projects, in the development of new technologies in the energy field, and in the construction of pipelines. They will serve a broad cross section of markets. They may enter fields of endeavor particularly conducive to maximizing profits through the application of their expertise and resources. At the bottom line, such companies will show above-average growth in sales and earnings. Such utilities will be recognized as industry leaders and as true growth companies.

In most instances, the quickest and best way to develop gas production capabilities is to acquire existing companies with specialized skills in that area. Gas production is becoming increasingly capital-intensive. Drilling offshore, in Alaska, or below 15,000 feet involves much technical know-how and plenty of money. Similarly, gas in tight formations, Devonian shale, and methane in coal seams will necessitate sophisticated technology and the investment of significant funds. The gas utility industry, with its strong marketing posture and sizeable capital base, is in a good position to enter this field. It is likely that many opportunities will arise over the next several years for making suitable acquisitions in the resource area and to expand operations through bidding on leases. The acquisition of conventional gas production capabilities should have top priority for any major gas utility that wishes to safeguard part of its resource base.

Mr. O.C. Davis, Chairman of Peoples Energy Corporation, told me that his company's experience led to a somewhat different approach. They found that the acquisition of existing companies in the gas exploration and production field was too costly a procedure. Instead, they hired a number of individuals with specialized know-how and built up their own organization.

If a gas utility discovers a major gas field not currently served by a pipeline, it may make sense to construct a pipeline to its own service area. An assured market can overcome many obstacles, including the raising of funds for pipeline construction.

For dynamic growth companies, financial planning is as important as industrial programming. Actually, the two are closely interrelated. Careful attention to cost-efficient operations, cash flows, return on capital, and increasing earnings per share of common stock are essential if investors are to be motivated to make additional funds available to help finance the growth of the gas utility industry.

Historically, the gas industry has been highly regulated. This reality has profoundly influenced managements and the manner in which they carried on their business. Getting along with the regulators has often had top priority. Generally speaking, this environment encouraged keeping a low profile and avoiding actions that might "rock the boat." Such an orientation will not be appropriate for dynamic growth companies.

If the gas industry is to meet its ambitious goals over the next twenty years, different management philosophies and regulatory policies will have to emerge. Decisive and aggressive leadership will be essential. Managements must be able to seize opportunities for profitable growth as they arise. Know-how regarding resources and production will become more important than they have been hitherto. Marketing will also have to become more dynamic. The gas industry may be compared to a sleeping giant that is in the process of awakening. Once it gets rolling, its performance may amaze many people, including those involved in running the industry.

Regulators will also have to develop new approaches to their tasks. The whole regulatory process should be streamlined and simplified, in order to facilitate the many management decisions that must be made in a dynamic enterprise.

Attracting talented management personnel must have top priority for any growth industry. The gas industry will provide many opportunities for able and ambitious individuals. Business and engineering school graduates will look with increasing favor on careers in the gas industry. Executives with outstanding performance records in other industries may be attracted to openings in gas companies. The management of gas utilities will increasingly reflect the new growth dynamics.

Growth is not just something that happens by itself; it has to be properly planned and nurtured. Planning by talented and ambitious people tends to develop a dynamic of its own; it has many self-fulfilling features.

When high caliber management is combined with adequate financial resources in a basic growth field, many opportunities for enlarged and diversified corporate activity will arise. New products and services will be offered. Acquisitions will be made in related areas. It is likely that the gas industry in the year 2000 will be far different from what it is now. Those companies which have the will to make the proper plans at this time are likely to emerge as the leaders of this new and exciting era.

25. The Hazards of Overregulation

Excessive regulation of the gas industry by federal, state, and local governments is a major impediment to increased utilization of gas for the national good. The gas industry is not well understood by many government officials. The adversary orientation that has characterized the approach of some regulators is often counterproductive, particularly if it is combined with political motivations. Regulatory procedures should be simplified and speeded up in order to expedite gas industry projects that can contribute to solving the energy problem.

The gas industry is regulated in virtually every facet of its operations. From the wellhead to the burner tip, gas is subject to a myriad of regulations by federal, state, and local governments. Often two or more agencies have overlapping or conflicting interests. The regulatory morass seriously interferes with the efficient operation of the gas industry.

The regulatory process is complicated by the fact that many government officials lack a basic understanding of the industry they are regulating. Some regulators appear to have a curious

118

prejudice against knowledge: they seem to confuse ignorance with impartiality. Experience has demonstrated the fallacy of this orientation. Sound judgment requires a thorough understanding of the broad picture as well as knowledge of the specific issues involved in a given case. Lectures, seminars, and courses dealing with the gas industry would be helpful in improving the situation.

In many respects, the relationship between government officials and the gas industry has been one of adversaries. Political motivations sometimes enter the picture. Some regulators may try to gain political advantages at the expense of the gas utilities they regulate by appealing to popular causes. Such an orientation can have harmful consequences to the gas utility industry and to the nation as a whole. For example, the gas pricing policies implemented by the Federal Power Commission between the 1950's and the 1970's played a major role in causing the energy crisis which still plagues this country. The record indicates that the public is not served well by the politicized adversary orientation of some regulatory officials.

The sheer magnitude of the regulatory case load is staggering. For example, the Federal Energy Regulatory Commission (FERC), which deals with many matters affecting the gas industry, has a backlog of some 20,000 cases.[1] As a result, applications for new energy facilities or other vital matters may take years to pass through the regulatory maze. The national interest requires fundamental changes in this situation if we are to deal effectively with the energy problem.

Leaders of the gas industry have made the following suggestions to expedite regulatory procedures:[2]

(1) The charters of regulatory agencies should be more carefully drafted.

(2) The number of regulations should be reduced.

(3) Regulations should be simplified.

(4) Regulators should be knowledgeable about the industry they are regulating.

(5) A fixed time schedule should be established for each case. Regulators should be obligated to make their decisions within that time frame.

(6) Flexible administrative procedures should be utilized.

(7) Much time could be saved if individual commissioners would handle cases from inception to final decision. In most cases it is not necessary for all members of the regulatory commission to make a joint decision.

(8) Congressional committees could fulfill a useful function by monitoring regulatory commissions and helping to eliminate abuses and unreasonable delays.

To sum up, the gas industry's contribution to solving the national energy problem can be greatly facilitated with a more realistic approach to regulation at all levels of government.

Sources:
[1]*George H. Lawrence and David J. Muchow, "The FERC's Case Load Management Problem,"* Public Utilities Fortnightly, *January 18, 1979.*
[2]*Ibid., "Regulatory Reform: Maybe a Pill with Nothing but Side Effects,"* Public Utilities Fortnightly, *March 27, 1980.*

26. Learning More About Natural Gas

The American people have much to gain from energy leadership by natural gas. It will hasten the achievement of energy independence. It will increase job opportunities. It will improve the balance of trade. It will reduce inflation. It will help clean the environment. It will make our nation more secure.

In view of these many benefits, natural gas deserves the active support of a broad cross section of the American people. By reading this book, you have already taken the first step in the right direction. Knowing the truth is an essential prerequisite to appropriate action. The process of understanding and intelligent behavior can be furthered by a program of lectures and workshops, such as the one outlined below.

(1) Introductory presentation about the role of gas in your community. The head of the local utility would be a good choice for this introduction, which would be followed by a question and answer period.

(2) Workshop on how you can save money and gas through improved equipment maintenance.

(3) Discussion about production of gas from shallow and deep conventional wells. If possible, individuals with practical experience in this field should be invited to make the presentation.

(4) Workshop on how homeowners can save money by having their furnaces converted from oil to gas.

(5) Panel discussion about new technologies for developing unconventional sources of gas, including coal and peat gasification. Personnel from the Institute of Gas Technology, Gas Research Institute, or the American Gas Association would be good resources for this topic.

(6) Workshop on new energy saving equipment, such as pulse combustion boilers and furnaces, gas-fired heat pumps and fuel cells.

(7) Presentation on how increased gas supplies can attract industry and create new jobs. A panel consisting of representatives from business, labor, and the gas utility would be the best format.

(8) Workshop on the use of methane for powering vehicles.

(9) Discussion on how gas can improve the environment. Representatives from environmental groups, scientists concerned with environmental matters, and government agencies should be invited to participate.

(10) Presentation on career opportunities in the gas industry. The local utility's personnel officer would be a good choice for this topic.

(11) Panel discussion on the proper roles of regulators and utilities in improving gas service to the community. The panel should include members of the regulatory commission and management representatives.

The lectures and workshops should be widely advertised and should be open to the general public. Students should be encouraged to participate. Educational institutions may consider giving credit to students who attend all presentations, and who prepare a report on their experience. Members of the press should also be invited. It would be desirable to broadcast these events over local radio stations.

The story of gas is well worth telling. On the basis of his own experience, the author believes that the more people know about gas, the better they will like it. The active involvement of the general public in this informational effort can play a vital role in facilitating the emergence of the gas age.

27. Impressive Leadership Qualifications

In my contacts with leaders of the gas industry, I was impressed by their dedication to customer service and to the national good. The gas industry has an outstanding record of promoting energy conservation. To assure long-term supplies, the gas utility industry is prepared to invest hundreds of billions of dollars. The gas industry leadership has offered its wholehearted cooperation in the effort to solve the nation's energy problem.

Gas companies serve over forty-seven million customers. This direct contact with consumers has profound implications. Leaders of gas companies try conscientiously to include the consumers' interests in their orientation and decisions. For example, it became apparent in the mid-1960's that unsound government policies in relation to wellhead gas prices were leading this country into a precarious situation. To deal with this predicament, the American Gas Association launched a national gas conservation program, long before energy conservation became fashionable.[1] Gas utility leaders also went on record as favoring more realistic gas pricing policies to provide adequate incentives for producers to step up the pace of exploration and production.

The gas utility industry has spent millions of dollars on energy conservation. In 1980 the Gas Research Institute will devote $16 million (32% of its research budget) to studying efficient utilization of gas.[1] The industry has taken an active role in the development of new, highly efficient home heating equipment, including pulse combustion devices, heat pumps, and fuel cells. Gas consumers were urged to install home insulation and to improve equipment maintenance. Significant results have already been achieved. Since 1973, gas energy consumption has been reduced about 16% due to consumer conservation efforts.[1]

The development of solar energy has been supported by many gas companies. For example, the Southern California Gas Company has an active research and development program in the solar field. In addition to promoting the use of passive and active solar devices, this utility is helping to finance the conversion of large numbers of homes from conventional water heating systems to solar units.[2]

Gas companies are prepared to spend $215 billion over the next two decades to assure their customers adequate supplies.[3] Most of these funds are dedicated to the development of long-range resources that would otherwise not be available, such as coal and peat gasification and new technologies for unconventional gas.

Leaders of the gas industry have expressed their willingness to work with the federal government in reducing U.S. vulnerability to oil import disruptions.[4] In line with this objective, they have proposed the use of gas in limited quantities along with substantial amounts of coal to facilitate conversion of electric utilities from imported oil to coal.[5] Such a joint endeavor between gas and coal would remove environmental obstacles to this program. If the federal government would implement the proposals made by the gas industry, U.S. dependence on imported oil would be sharply reduced in a relatively short time.

It is noteworthy that the leaders of the gas industry are genuinely dedicated to helping the U.S.A. solve the energy problem. They are prepared to place the national good ahead of any selfish motives. They are not trying to make quick or exorbitant profits at the expense of their customers. They are willing to cooperate with all levels of government and with other suppliers of energy to further the goal of greater self-sufficiency for the nation's benefit.

Leadership has many dimensions. In addition to demonstrating practical problem-solving abilities, leaders must show awareness of the broader social framework within which they function. The energy crisis poses major practical and ethical challenges to the political and business leaders of this country. To gain the confidence of their fellow men, leaders must combine moral rectitude with sound decisions. The leadership of the gas industry is well positioned to serve this nation in its quest for a better future.

Sources
[1] *"Energy Efficiency and the Natural Gas Industry," American Gas Association, February 1980.*
[2] *"The Role of Gas in Southern California," Southern California Gas Company, May 1980.*
[3] *"A Forecast of Capital Requirements of the U.S. Gas Utility Industry to the Year 2000," Policy Evaluation & Analysis Group, American Gas Association, April 20, 1979.*
[4] *"Written Comments of the American Gas Association to the U.S. Department of Energy to Assist in Developing a Plan for Substantially Reducing by 1985 U.S. Vulnerability to an Oil Import Disruption," by George H. Lawrence, Michael I. German, and Donald J. Schellhardt, May 30, 1980.*
[5] *"Analysis of Select Gas Use in Utility Coal Conversion for Limiting Sulfur Emissions," Policy Evaluation & Analysis Group, American Gas Association, April 29, 1980.*

28. Clarifying Questions and Answers

The question and answer format can be useful in clarifying some of the ideas presented in the book. This approach can also elaborate upon issues and give additional information. Finally, it provides a forum for informal comments and expressions of personal opinions.

(1) How did you select the sources of information for this book?

Answer: I was influenced by the availability of up-to-date and comprehensive data as well as by my assessment of the sources' qualifications. It is apparent from the many references that I utilized academic, governmental, and business sources, with particular emphasis on the latter.

Let me explain why I paid special attention to gas industry sources. Professionals in the gas industry have spent many years of full-time service to develop their expertise. Their information is well seasoned with practical experience. Their leadership status depends on a thorough understanding of current and future trends affecting the industry. They must be familiar with all facets of the business, including scientific, technical, economic, and financial aspects. If they want to further their careers, they must demonstrate sound judgment and the ability to achieve results.

The information I received from executives and from associations serving the gas industry reflects this broadly based approach. It has the additional advantage of providing an insight into possible future happenings, because ideas and actions are closely linked in the orientation of industrial decision makers.

(2) What is the difference between methane and natural gas?

Answer: Methane is the main ingredient of natural gas. The latter may also contain other types of hydrocarbon, such as ethane, propane, and butane. Some carbon dioxide, nitrogen, and other trace elements may be present in natural gas. Generally, methane makes up over 90% of pipeline quality natural gas. In this book, the terms "natural gas" and "methane" are used interchangeably.

(3) How does gas provide energy?

Answer: When methane burns by combining with oxygen, it gives off heat energy. This heat can be used to cook, to run a furnace or appliance, to power an industrial plant, or to generate electricity. Let us look more closely at this process. When you turn on the gas on your range, three factors combine to start the combustion process: methane, oxygen, and a source of ignition, such as a pilot light or an electronic igniter. Under these circumstances, each methane molecule combines with two oxygen molecules to form water vapor and carbon dioxide. Using the shorthand formulations developed by chemists, this happening is described as follows: CH_4 (methane) plus $2O_2$ (oxygen) $\rightarrow 2H_2O$ (water) plus CO_2 (carbon dioxide). It is noteworthy that the number of atoms on both sides of the equation is the same. The rearrangement of atoms into different molecules releases energy.

(4) What happens to the water and carbon dioxide that result from the burning of gas?

128

Answer: Both the water molecules and the carbon dioxide enter the surrounding air. The water vapor will subsequently cool off and condense. Much of the carbon dioxide may eventually be taken up by plants. The latter use the carbon as a basic ingredient of their substance and release the oxygen into the air. When plants have completed their life cycle, they may be converted into methane by microorganisms.

(5) What causes the smell if gas leaks without burning?

Answer: Methane is odorless. The smell from leaking gas is due to an odorant that has been added as a safety measure. Whenever you smell the odorant, it means that some gas is leaking. If you have any problem with leaking gas, you should contact your gas utility or fire department without delay.

(6) Where and when was natural gas first discovered?

Answer: There is some evidence that the ancient Chinese used gas as a fuel some 3,000 years ago. They utilized bamboo tubes as pipes. One of their applications for this fuel was to evaporate brine to make salt ("The Story of Gas," A.G.A. Monthly, August 1975). In the U.S., credit for starting the commercial use of natural gas is given to Fredonia, New York, about 40 miles from Buffalo. Gas was discovered in that area in 1821 (*Ibid.,* October 1975).

(7) Where can gas be found?

Answer: Gas can be situated almost anywhere. Near the earth's surface, places containing organic debris covered by earth or water, such as landfills and swamps, are sources of gas. As one goes deeper into the earth, sedimentary basins capped by rocks or salt domes in which gas can accumulate, are logical places to look for gas. Wherever there is oil, there is generally also gas. Moreover, gas can exist in places unsuitable for oil, such as deep geological formations with high temperatures. The science of geology has made great advances in mapping out potential gas-containing areas. However, the ultimate answer can only be found by means of drilling. Time and gain, gas has been discovered in unpromising areas. On the other hand, locations with supposedly good potential may result in dry holes. Exploration for gas is a challenging, high risk business.

(8) Why are you so optimistic about the prospects for gas below 15,000 feet?

Answer: Economic, geological, and technological factors have combined to make the outlook for deep gas very promising. The removal of price controls on gas from below 15,000 feet has provided the economic incentives for drilling to great depths. My background in finance makes me pay particular attention to the economics of drilling. It is realistic to assume that businessmen will spend $5 to $15 million for drilling a well only if they believe that the price of gas they wish to discover is adequate to compensate them for the risks involved. It should be remembered that drillers for gas are businessmen who are guided in their decisions by the profit motive. All too often in the past our government officials seem to have overlooked this basic reality in dealing with energy matters. In any case, the current economic environment for deep drilling is favorable.

Geology favors the discovery of prolific gas fields in suitable locations at great depths. The generation of gas from organic sediments combines with the conversion of oil into gas at high temperatures to set the stage for vast gas accumulations. In the brief period since deep drilling was started, a number of giant gas fields have already been discovered. Dr. John M. Hunt of the Woods Hole Oceanographic Institution makes a convincing case for the probable exisence of many locations where these favorable conditions for gas may exist. The Tuscaloosa Trend, Anadarko Basin, and Rocky Mountain Overthrust Belt may well be the forerunners of many more giant deep gas fields.

The technological developments that have made possible

130

commercial drilling for deep gas are quite impressive. In many respects, the challenges involved in drilling four to six miles into the earth are as great as those characteristic of space exploration. The scientists and engineers who developed the specialized equipment are among the unsung heroes of our time. It is truly remarkable that wells going 20,000 or more feet into the earth are currently being drilled in an almost routine fashion.

The saying "necessity is the mother of invention" is applicable to deep gas drilling. The development of the sintered bauxite proppant by Exxon illustrates this point. Because of the enormous pressures existing at great depths, openings (fractures) through which gas flows to the surface have a tendency to shrink and may even close altogether. It is essential to counteract this tendency by inserting a material which keeps the fracture open. In shallow locations, coarse sand can be used for this purpose, but at great depths sand is pulverized. To deal with the problem, Exxon scientists tested a great many materials and discovered that sintered bauxite was best suited for the purpose. Without this super-strong new proppant, drilling for deep gas would probably have had only limited appeal. With the help of this proppant, many deep wells produce 20 million cubic feet of gas per day or more. On that basis, deep wells make good economic sense.

(9) Do you think that deep gas is more promising than gas from shallower locations?

Answer: There is undoubtedly much gas that remains to be discovered in locations nearer the surface than 15,000 feet. If the government would open up more of its 2 billion acre holdings to gas exploration I would become more optimistic about shallow gas. The gradual removal of price controls on new gas will also help. As I see it, if we want gas to become the new energy leader, we need to develop all of our resources, shallow and deep.

(10) How do you feel about the commercial prospects for unconventional gas?

Answer: Dr. Henry Kent, Director of the Potential Gas Committee at the Colorado School of Mines, pointed out to me that the distinction between "conventional" and "unconventional" gas is by no means clearcut. Technological developments can narrow or even remove the gaps between these categories. For example, the discovery of new fracturing techniques for recovering gas in tight sandstone formations may add many trillions of cubic feet of gas to available supplies. It does not really matter how we categorize this gas. Methane is methane, no matter where it comes from.

I am in favor of active research and development programs for all the so-called unconventional sources of gas that show promise. As soon as technology permits economic operations, these gas resources should be brought on stream to add to supplies.

(11) If the outlook for conventional and unconventional gas is so promising, do we still need gasification of coal and peat?

Answer: The answer is yes. It is important to remember that, if the industry accepts the full challenge of energy leadership, it should try to reach a level of production approximating 40 trillion cubic feet of gas by the year 2000. Even though I am optimistic about conventional sources of gas, I feel we need a broad, diversified base from which gas can be supplied. Moreover, while the initial cost of gasification may be higher than conventional gas, technological developments are likely to lower future gasification costs. The byproducts of gasification may also contribute significantly to the economic soundness of this approach. Long-term planning is essential for a realistic energy program. Coal and peat are likely to supply large quantities of gas for many decades, if not for several centuries.

132

(12) How do you feel about the economic soundness of ocean energy farms?

Answer: The work that has been done thus far on giant sea kelp looks promising. The concept of using nature's processes for growing plants and converting them into methane makes good sense. Long before the pessimists predict an end to our fossil fuel supplies, we will have as much gas from renewable sources as we need. I was so intrigued with this concept that I originally thought of entitling this book, *GAS ENERGY FOREVER!*

(13) Wouldn't it be simpler to grow plants on land for methane production?

Answer: Areas available for energy production on land are much more limited than those in oceans. Most of the land that can be used for plant growth is essential for the production of food and fiber. The ocean has several major advantages: (a) virtually unlimited space; (b) readily available plant food (no fertilizer is necessary); (c) existing plants seem ready-made for optimum methane production; and (d) the possibility of adding to the world's supply of high-protein seafood along with the methane. The world's oceans could supply all the energy human beings need in the foreseeable future. International cooperation on ocean farming could lead to a better world for all of mankind. I believe we should pursue this enterprise with enthusiasm.

(14) In your critique of the gas industry's capital spending plans you indicate preference for additional spending on conventional gas supplies. Does this imply that less money should be spent on unconventional sources of gas?

Answer: That is not my intention. I would like to see the gas utility industry spend considerably more on conventional gas exploration and development than the $85 billion currently programmed. A doubling of that figure would probably make good sense. On that basis, the $130 billion to be spent on supplemental sources of gas would be in good balance with the overall spending plans. It should be noted that the implementation of my suggestion would raise overall capital expenditures for the next two decades to almost $400 billion. It would also raise the targeted market supplies of gas from 30 trillion cubic feet to 40 trillion cubic feet by the year 2000.

(15) How important is marketing to the achievement of gas energy leadership?

Answer: Marketing is of the utmost importance if gas is to achieve its objectives. The gas shortages that have plagued the industry in the 1970's may have given the misleading impression that marketing is secondary to assuring adequate supplies. This orientation is not valid under present and foreseeable circumstances. Supplies of gas are becoming increasingly plentiful. Some U.S. gas wells have been shut in for lack of markets and Canadian supplies available for export are not being fully utilized. The gas industry will have to make an aggressive marketing effort to keep exploration and production of gas at a high pitch. All restrictions on marketing of gas imposed by the federal government should be lifted without further delay.

(16) Do you consider increased production of gas more important than conservation?

Answer: I believe strongly in prudent use of natural resources. We should do everything possible to avoid wasting energy. Furnaces and other gas-using equipment should be put into optimum working condition through proper maintenance. Most people could probably readily save ten percent or more on monthly gas bills if they followed this procedure.

I have devoted more of this book to gas resources and potential supplies because that topic has been strangely neglected by many writers and by virtually all government spokesmen. It appears as if most people in this country have accepted the fallacious notion that we are about to run out of gas. Some government officials seems to have a vested interest in promoting the idea that our gas resources are shrinking rapidly. That assessment is the only basis on which their unsound policies could have been justified. If our gas resources turn out to be plentiful, as I am confident they will, one can readily conclude that the government's policies have been wrong.

(17) What do you think has been wrong with the federal government's energy policies?

Answer: Between 1954 and 1978, the federal government interfered with the laws of supply and demand by putting artificially low prices on gas at the wellhead. These prices stimulated demand for gas, while removing the incentive for producers to add new supplies. The inevitable outcome was the experience of gas shortages in the 1970's. If anybody had deliberately tried to create an energy crisis, he could not have done a better job than the people in government responsible for our energy policies.

(18) How do you account for these unsound government policies?

Answer: Politicians generally try to curry favor with the voters by doing them supposed favors. Low utility prices are popular and therefore politically attractive. The fact that these shortsighted policies have disastrous long-term consequences is often overlooked.

Another problem is that most bureaucrats have little or no understanding of the field they are regulating. Mr. B.Z. Kastler, chairman and president of Mountain Fuel Supply Company, expressed this reality in succinct words: "Behind every 'doer' in our society is a veritable host of critics, reviewers and second-guessers. Most are bureaucrats, selected for their 'impartiality,' which, translated into modern vernacular, means their total lack of knowledge and experience in the field they are regulating." (presentation at the Institute of Gas Technology, November 15, 1979).

Most academicians and the mass media have not been very helpful, either. Many of them favored the political philosophy underlying these unsound practices, or were incapable of analyzing the situation realistically. A combination of anti-business sentiment and ignorance of basic economics can go far in explaining the blindness to economic realities that has been so widespread in our society during the past several decades.

(19) Do you see any hopeful signs that this orientation is changing?

Answer: I believe the realities of the energy crisis have begun to make their mark. Economic realism is on the increase. The passage of the Natural Gas Policy Act of 1978 was a step in the right direction. The 1980 platforms of both political parties contain planks that favor increased use of government lands for energy exploration and production. We are beginning to see a movement away from government controls over prices, production, and marketing of gas. I hope this trend will accelerate.

(20) Isn't there a danger that decontrolled gas prices will go too high?

Answer: On the basis of historical experience, fear of a free market is unfounded. When prices are determined by demand and supply, they stabilize at levels that are realistic in terms of the best interests of the parties involved. Producers will have sufficient incentive to seek and produce enough gas to meet the demand; consumers will have sufficient incentive to purchase the gas. If the prices are high because of temporary supply shortages, producers will have the incentive to increase output, while consumers will be motivated to reduce consumption. Both of these factors will result in reducing pressures on prices. The free market is a remarkable medium for optimizing realistic economic behavior.

Fear of the free market, like all fears of reality, are based on unsound assumptions. We have experienced price controls on gas by government fiat for such a long time that many people, particularly those in government, have an irrational concern about what would happen if these controls were lifted. Well, the heavens won't fall down. Furthermore, the lifting of controls will remove the counterproductive results which we have experienced during the past several years.

(21) How do you think the oil industry will respond to the challenge from gas?

Answer: Leadership in any field should be a function of merit and performance. Under the conditions prevailing now and in the foreseeable future, gas is in a better position to provide additional supplies of energy from domestic sources than oil. The oil companies will actually benefit from a broadened market for gas, as they are the main gas producers. As utilities enter the field of gas production, they will provide more competition to oil companies, though they will also continue to be major customers. As far as I am concerned, increased competition is in the best interests of the country.

(22) Won't the oil companies resist losing the residual fuel market to gas?

Answer: Rational economic behavior is based on optimizing profits, not on illusory notions about holding on to markets as an end in itself. Residual fuel oil is a leftover from the refining process that is sold at much lower prices than transportation fuels. According to the Purvin & Gertz study ("An Analysis of Potential for Upgrading Domestic Refining Capacity," May 1980), most of this residual fuel oil could be upgraded into gasoline or diesel fuel, thus greatly increasing the return on investment to the oil companies. It seems to make eminent sense for oil refiners to follow this procedure. The decision by Ashland Oil to move in this direction confirms this assessment.

(23) How will the coal industry react to the new energy leadership by gas?

Answer: The coal industry has much to gain from such a development. Gas utilities will become increasingly important customers of coal. If all the coal gasification projects currently planned are implemented, the gas industry will probably be the largest single market for coal by the end of this century. It is noteworthy that much of this coal would not have found any other markets, because of low quality or high levels of pollutants. Moreover, gas will facilitate the use of coal by electric utilities. The addition of modest quantities of gas will enable electric utilities to use large amounts of coal without harming the environment significantly more than had been the case with oil.

(24) How do those concerned about the environment respond to gas?

Answer: Scientists and others who are concerned about the environment welcome gas as a desirable fuel. Gas is relatively nonpolluting. One can make a convincing case that gas is the most natural way for utilizing solar energy. It is important that those concerned about the environment support not only the increased use of gas, but also active exploration and development programs.

(25) When do you think gas will emerge as the new energy leader?

Answer: Leadership is as much a state of mind as it is a statistically verifiable condition. As far as I am concerned, gas is the heir apparent to energy leadership. Gas has all the qualifications for leadership. It is likely that an increasing number of people will share my views as they familiarize themselves with the facts. The gas industry is in a position to increase its services to an ever larger number of customers. In the process, it will replace oil in many areas, particularly in supplying heat and power for stationary applications in homes, commercial establishments, and industries. Methane may also find increased acceptance for powering vehicles, as feedstock for the manufacture of chemicals and plastics, and for other applications. Dynamic managements of gas utilities will find the sources of supply, the markets, and the financial resources that will confirm gas as the new energy leader. This development will benefit all of us as consumers and as citizens of a society moving in the direction of greater energy independence.

(26) Has reliance on gas imports contributed to the energy crisis in the United States?

Answer: Between 1955 and 1979, imports of natural gas averaged less than five percent of total U.S. supplies. Most of these imports came from Canada. A realistic assessment of the facts indicates that imports of natural gas had nothing to do with causing the energy crisis of the 1970's.

(27) Are U.S. government officials aware of the important role gas can play in helping to resolve the energy crisis and in creating a more wholesome economy?

Answer: The gas industry has made an active effort to share its ideas with government officials. Among the many communications that took place was a recent letter by Mr. George H. Lawrence, president of the American Gas Association. Dated August 29, 1980, it was addressed jointly to Mr. G. William Miller, Secretary of the Treasury, Mr. Charles Schultze, Chairman of the Council of Economic Advisers, and Ms. Anne Wexler, Assistant to the President. This letter contained the following remarks:

"In my letter of August 21, 1980 to you on the role which gas and coal can play together in displacing oil imports, I discussed the progress to date the gas industry has made in backing oil out of non-transportation energy markets, and I emphasized that the gas industry contributions could be much greater. We have just completed an Energy Analysis entitled, 'Potential Substitution of Oil with Gas and Coal in Non-Transportation Uses,' copy attached, which details how much more oil can be displaced and where that oil is presently being consumed. The dramatic, but highly encouraging result is that this analysis clearly shows where oil consumption, up to 5.5 million barrels per day, can be displaced with domestic fuels."

The data on oil displacements referred to in the letter are presented in chapter 22 of this book (see table entitled "Oil Use Replaceable by Gas and Coal"). It is noteworthy that 5.5 million barrels of oil a day amounts to about 2 billion barrels on an annual basis. At the recent price of $30 a barrel, this replacement of oil by gas and coal would save the U.S. about $60 billion in direct annual import costs. If Professor Rod Lemon's calculations on the "external costs" of imported oil were included (see chapter 13 for details), total savings to the nation might range from $120 to $150 billion per annum.

Such a development would have the following implications for the national economy:

(a) The balance of payments would shift dramatically from large deficits to sizeable surpluses.

(b) The U.S. dollar would become one of the strongest currencies in the world.

(c) Inflation rates would be sharply reduced.

(d) The U.S. economy would be revitalized, led by domestic energy industries.

(e) Employment would rise significantly and unemployment would decline.

(f) The costs involved in stockpiling oil and in safeguarding oil supply lines halfway around the world would be minimized. Overall, the U.S. would be in a better economic position than it has experienced for years. The gas industry has gone on record with government officials that it is prepared to play a vital role in helping to achieve this desirable state of affairs.

29. Some Vital Gas Statistics

U.S. gas production in 1979: 19.9 trillion cubic feet.

Proved reserves of natural gas in the U.S. at year-end 1979: 195 trillion cubic feet.

Number of customers served by the gas utility industry in 1979: 46.8 million. This total included 43.2 million residences; 3.4 million commercial establishments; and 193,000 industrial users.

Gas utility and transmission pipelines totaled 1,013,000 miles.

Storage capacity for natural gas was 7.3 trillion cubic feet.

The gas utility industry employed 208,000 people in 1978.

Industries and commercial establishments depending primarily on natural gas as a source of energy employed 25 million workers with an annual payroll of $325 billion in 1978.

Sales of central gas heating equipment in 1978 totaled over 1.7 million units.

Residential customers consume 85% of their gas for heating purposes.

Total U.S. energy consumption in 1978 was 78.1 quadrillion British thermal units.

Percentage breakdown of U.S. energy consumption in 1978: petroleum (oil), 48.6%; natural gas, 25.4%; coal, 18.2%; hydro and nuclear, 7.8%.

Domestic U.S. energy production, percentage breakdown: natural gas, 35%; oil, 30%; coal, 25%; hydro and nuclear, 10%.

Oil imports represent over 40% of total U.S. oil consumption. Such imports are estimated to cost $80–90 billion in 1980.

6,000 cubic feet of natural gas equal one barrel of oil.

1,000 cubic feet of natural gas contain one million British thermal units of energy.

One trillion cubic feet of gas can replace 500,000 barrels of oil a day for a year.

A six trillion cubic feet gas field is equivalent to a one billion barrel oil field.

Glossary

Abiogenic theory of methane's origin: the hypothesis that methane was one of the original building blocks of the earth, with much of it being trapped in the planet's interior. According to this hypothesis, most of the methane found in the earth's crust has moved up from the lower regions. Therefore, it is claimed that methane comes primarily from non-biological sources. This hypothesis remains to be proven; it is not widely accepted by scientists.

A.G.A.: American Gas Association.

American Gas Association: the trade association of the gas utility industry. About three hundred members belong to this organization.

Anadarko Basin: a large area covering parts of Oklahoma and Texas which is a major source of natural gas from both shallow and deep formations.

Anaerobic microorganisms: microscopic forms of life that thrive in an environment not containing air or oxygen.

Aquifer: a rock formation containing water.

145

Biodigester: an airtight container in which anaerobic microorganisms convert organic debris into methane.

Biomass: any organic material, particularly plants and their residue, which can be utilized to produce methane.

British thermal unit: the amount of heat required to raise the temperature of one pound of water one degree Fahrenheit.

Btu: British thermal unit.

Captive sources of natural gas: supplies that are owned by gas pipeline companies or by gas utilities.

Carbon dioxide: a chemical molecule consisting of one carbon atom and two oxygen atoms (CO_2). Plants take in carbon dioxide, retain the carbon, and release the oxygen. Carbon dioxide is an important byproduct of coal gasification; it can be used for enhanced oil recovery, to increase oil production.

Conventional sources of methane: natural gas that is situated in reservoirs which can be reached with conventional drilling procedures. Such gas may be in shallow or deep wells. It may be associated with oil or it may occur in reservoirs not containing oil.

Crude oil: a liquid mixture of hydrocarbons found in underground reservoirs. In its original form, crude oil cannot be utilized as a source of energy. It must first be processed (refined) to be made into useable products, such as gasoline, diesel fuel, kerosene, or fuel oil.

Crust of the earth: the top layer of the earth, approximately twenty-five miles deep.

Devonian shale: gas-containing rocks which were formed in the Devonian period, about four hundred million years ago. These rocks release gas only slowly; special procedures, such as the use of explosives or hydraulic fracturing, are necessary to speed up gas recovery. Most of the Devonian shale in the U.S. is situated in the Appalachian area.

Eastern Overthrust Belt: the region in the eastern U.S.,

ranging from New York to Alabama, that appears to have promising potential for becoming a major source of natural gas. See "Overthrust Belt" for additional information.

External economic costs: those costs of economic transactions which are born by society as a whole, not by the parties directly involved as buyers and sellers. For example, uncontrolled air pollution is an external cost. When companies install air pollution control equipment, the external cost is transformed into an internal one, that is, it becomes part of the cost of doing business. Professor Rod Lemon has devoted considerable effort to determining the external costs of imported oil. See chapter 13 for details.

Flaring of gas: setting gas on fire as it escapes into the atmosphere. This procedure is sometimes used in connection with oil production and refining.

Fuel cells: devices that use natural gas to generate electric power at the point of end use, such as apartment houses, commercial buildings, and industrial plants. Gas-fired fuel cells are expected to be introduced commercially in the mid-1980's. They can contribute significantly to saving energy.

Gasification: the conversion into gas of such fossil fuels as coal, peat, and oil shale. The procedure involves a combination of heat and chemical processing.

Geopressured methane: gas dissolved in hot brine (salt water) under great pressure at deep locations.

Heat pump: a device for producing heat in the winter and cooling in the summer. Several types of gas-powered heat pumps are being developed. Commercial introductions are expected in the mid-1980's. Heat pumps help to save energy.

Heavy hydrocarbons: compounds of hydrogen and carbon in which the carbon (which is heavier than the hydrogen) is more predominant. Heavy hydrocarbons are either liquid or semi-

147

solid. In contrast, light hydrocarbons, such as methane, tend to be gaseous.

Hydrates: solid compounds of methane and water, which are often found in permafrost areas of the Arctic and in some deep ocean locations.

Hydraulic fracturing: the use of water or other liquids under great pressure to force openings into underground formations containing natural gas.

Hydrocarbons: molecules made up of hydrogen and carbon. Methane (CH_4) is a hydrocarbon normally in gaseous form.

In situ coal gasification: the use of controlled fire (heat) in a coal seam to produce methane. This procedure may be contrasted with conventional coal gasification, which involves mining the coal and transporting it to surface facilities, where it is gasified.

Landfill: a garbage dump that is covered with earth to provide more sanitary conditions. By keeping air (oxygen) away from the garbage, landfills make it possible for anaerobic microorganisms to convert much of the garbage into methane.

Lignite: brown coal, which will be used in the first commercial coal gasification project currently being constructed in North Dakota.

Liquefied natural gas (LNG): when methane is cooled to $-260°$ Fahrenheit, it shrinks to 1/600 of its volume and becomes liquid. LNG is placed in insulated containers, which keep it liquid at atmospheric pressure. This procedure makes possible the transporting and storing of large amounts of energy on an economical basis.

LNG: liquefied natural gas.

Mantle of the earth: the layer between the crust and the core. The mantle is estimated to be about 1,500 miles deep.

Methane: the hydrocarbon molecule consisting of one carbon atom and four hydrogen atoms (CH_4). Methane is the principal ingredient of natural gas.

Natural gas: a gas found in nature which consists primarily of methane. Other components which may also be present include such hydrocarbons as ethane, propane, and butane. In addition, natural gas may contain carbon dioxide, nitrogen, helium, and sulfur. The presence of sulfur turns the gas sour. Sulfur has to be removed before the gas can be put into pipelines.

Natural Gas Policy Act of 1978: the legislation by the federal government which may be credited with having created a more favorable environment for the exploration and production of natural gas.

Nitrogen oxides: molecules containing nitrogen and oxygen. Such molecules are formed whenever combustion takes place in the presence of air. The latter contains nitrogen as well as oxygen.

Oil: a liquid mixture of hydrocarbons.

Overthrust belt: an area which has been formed by the collision of subterranean rock formations, resulting in natural gas-containing sediments being relocated in deeper regions. See footnote in chapter 4 for additional information.

Peat: plant materials that have been partially converted into carbon, usually found in wet areas, such as bogs. Peat can be used as a soil conditioner, fertilizer, and fuel.

Phillips Petroleum Case: the 1954 U.S. Supreme Court decision which mandated that the Federal Power Commission should regulate the wellhead price of gas paid by interstate pipeline systems to independent producers. On the basis of this decision, the Federal Power Commission kept prices at levels that were too low to motivate producers to explore for adequate amounts of new gas supplies. The gas shortage of the 1970's was an outgrowth of this policy.

Photosynthesis: the process whereby plants convert sunlight into chemical energy by means of their green chlorophyll-containing cells. This procedure enables plants to utilize inorganic substances, particularly carbon dioxide and water, for conversion into organic compounds, and to release oxygen into the atmosphere. All life on earth, including that of human beings, depends on this process.

Potential Gas Committee: a group of about 140 experts on gas, including geologists, oil and gas engineers, specialists in exploration, production and other technologies, representatives of government agencies, pipelines, and gas distribution companies, and independent consultants. This organization, which is headquartered at the Colorado School of Mines, prepares periodic estimates of gas resources and reserves in the U.S.

Potential gas resources: quantities of natural gas in conventional reservoirs that are recoverable with existing technology, but that have not as yet been found through drilling. Potential gas resources may be divided into the following categories: (1) probable; (2) possible; and (3) speculative. "Probable" potential gas is the supply that is expected to result from the extension of existing gas fields. "Possible" potential gas is the supply expected from the discovery of new fields in geological formations that have already proved productive elsewhere. "Speculative" potential gas is the supply that is expected to result from drilling in new areas that have not produced gas before, but that appear to have promising prospects.

Primary recovery: the amount of gas or oil that can be recovered from a well on the basis of its own power, without having to apply external sources of energy.

Proppants: materials that are used to facilitate production of gas by reenforcing the openings through which the gas is extracted. The most commonly used proppant material for shallow wells is coarse sand. The latter is not strong enough to withstand pressures at great depths. For deep gas recovery, sintered bauxite, which is very strong, is the most effective proppant currently available.

Proved reserves: known quantities of gas that can be recovered from existing reservoirs. Geologists use information obtained from drilling, production-pressure relationships, and other data to determine the volumes of gas that may be recovered from these sources. Proved reserve figures are usually fairly accurate and reliable.

Pulse combustion: a procedure whereby heat is generated by gas burning in a sealed chamber. This approach avoids the loss of heat that normally goes up the chimney; in fact, pulse combustion furnaces and boilers do not require chimneys. The efficiency of pulse combustion devices exceeds 90%.

Pyrolysis: the use of heat to convert organic materials into methane.

Renewable energy sources: energy from sources that can be relied upon to recur in the future. The sun is the prime example of a renewable energy source. In contrast, fossil fuels, such as coal, oil, and natural gas, are usually considered to be nonrenewable sources of energy. However, as has been shown in this book, methane (natural gas) can be produced from plant materials. It is therefore appropriate to consider methane a renewable source of energy.

Residual fuel oil: the oil that is left over after the conventional refining process has been completed. Residual fuel oil is usually sold at low prices to electric utilities or industrial customers. Technologies have been developed to upgrade the refining process in such a way as to maximize production of gasoline and diesel fuel and to minimize the amount of residual fuel oil. As a result, the income of oil refiners can be substantially increased.

Rocky Mountain Overthrust Belt: a vast area stretching from Canada to Mexico in the Rocky Mountain region that contains large amounts of natural gas. See "overthrust belt" for additional information.

Sediments: accumulations of organic materials that settle at the bottom of water bodies, such as lakes or ocean areas adjacent to rivers. Organic sediments decompose through the action of microorganisms and/or heat, forming methane as one of the end products.

Solar energy: the sun generates energy through the fusion of atoms. This energy reaches the earth in the form of radiation, such as light waves. Plants utilize solar energy to promote their own growth by means of photosynthesis. When plants have completed their life cycle, they decompose. Methane may be formed as a byproduct of plant decomposition. In this manner, methane is a renewable form of solar energy.

Sour gas: natural gas which contains sulfur. The sulfur is removed before the gas enters interstate pipelines.

Stationary applications of gas energy: the use of gas to provide heat and power in homes, commercial buildings, and factories. This category may be contrasted with applications involving movement, such as transportation. Natural gas is a superior fuel for most stationary applications.

Sweet gas: natural gas that is free from sulfur.

Tight sands: sandstone formations which lock gas so tightly that special procedures are necessary to release gas at economic rates. The largest known accumulations of gas in tight sands are located in the Rocky Mountain areas.

Trillion: one thousand billion, or 1,000,000,000,000.

Tuscaloosa Trend: a 200-mile long, 30-mile wide geological formation in southern Louisiana, which appears to contain large quantities of deep gas. Estimates of potential gas resources in this area run as high as sixty trillion cubic feet.

UNITAR: United Nations Institute for Training and Research, a specialized agency of the United Nations. One of the functions of this agency is to sponsor conferences on subjects of worldwide interest, such as energy. Participants in such conferences include experts from many countries.

Venting of gas: letting natural gas escape into the air. This procedure is widely used in connection with methane contained in coal seams.

Wellhead: the well from which gas is extracted. Also, the source of gas.

About the Author

Ernest J. Oppenheimer, Ph.D., the author of this book, has followed the energy problem ever since it became acute in 1973. He devoted the past two years to intensive research on that subject. The first product of this research was the book *A Realistic Approach to U.S. Energy Independence,* which focused primarily on oil resources and presented a critique of government policies. The interest in gas was stimulated by these earlier studies. The current book is an outgrowth of the author's conviction that gas is going to be an increasingly important source of energy in the U.S.A.

Previously, Dr. Oppenheimer had worked on inflation. His book *The Inflation Swindle* was published by Prentice-Hall in 1977. He also wrote a booklet entitled *What You Should Know About Inflation.*

His fifteen years experience doing research and consulting work in the investment banking field provided useful background for dealing with the energy problem. His work revolved largely around technological companies, thus requiring an understanding of both science and finance. He had to develop practical solutions to complex problems in the investment field; this skill has direct applicability to the energy area.

He received the doctor of philosophy degree in international relations from the University of Chicago. His studies included political science, economics, history, social psychology, international law and diplomacy.

Dr. Oppenheimer has been interviewed on television and radio talk shows and has lectured to business groups and to university audiences.

ARE YOU INTERESTED IN ADDITIONAL COPIES OF THIS BOOK?

If you feel that this book has given you a better perspective on natural gas and its role in the overall energy situation, you may wish to make it available to others. An informed public is the best safeguard for sound energy policies.

The price is $7.50 per copy. For each five copies of the book you order at $7.50 each, you will receive one free copy. If you purchase one hundred copies or more, the price is reduced to $5.00 per copy.

BOOK ON U.S. ENERGY INDEPENDENCE

Dr. Oppenheimer has written a book entitled *A Realistic Approach to U.S. Energy Independence,* which may be considered a companion piece to the book on natural gas. The following quote expressed the book's basic thesis: "There is plenty of oil and gas in the U.S.A. The main shortage appears to be one of common sense and economic realism on the part of politicians and bureaucrats." This original trade paperback contains 92 pages. The price is $5.00 per copy. For quantity discounts, please contact the publisher.

BOOKLET ON INFLATION

Dr. Oppenheimer's booklet *What You Should Know About Inflation* has been widely praised for its clear appraisal of the inflation problem. Over 60,000 copies have been sold. This forty page booklet can be purchased for $2.50 a copy. For quantity discounts, please contact the publisher.

ORDER FORM

(1) *Natural Gas: The New Energy Leader*
   Number of copies——— Price per copy———
                                    Total———
(2) *A Realistic Approach to U.S. Energy Independence*
   Number of copies——— Price per copy———
                                    Total———
(3) *What You Should Know About Inflation*
   Number of copies——— Price per copy———
                                    Total———
   Residents of New York State add sales tax   ———
                                    Total cost———

All prices include regular delivery. Special charges, such as express mail and air shipments overseas, must be paid by the customer. Please enclose payments with mail orders. Make checks payable to Pen and Podium Productions. Send your order to:

Pen and Podium Productions
40 Central Park South
New York, N.Y. 10019

Please print or type your name and address below.

Name———————————————————

Company————————————————

Street—————————————————

City, State, Zip———————————————